ROCKY
MOUNTAIN DIVIDE

ʌʌʌʌʌ

ROCKY
MOUNTAIN
DIVIDE *Selling and Saving the West*

JOHN B. WRIGHT

Austin UNIVERSITY OF TEXAS PRESS

Requests for permission to reproduce material from this work should be sent to Permissions, University of Texas Press, Box 7819, Austin, TX 78713-7819.

∞ The paper used in this publication meets the minimum requirements of American National Standard for Information Sciences—Permanence of Paper for Printed Library Materials, ANSI Z39.48-1984.

Library of Congress Cataloging-in-Publication Data

Wright, John B. (John Burghardt), date.
    Rocky Mountain divide : selling and saving the West / John B. Wright.—
1st ed.
        p.   cm.
        Outgrowth of author's thesis (Ph. D.)—University of California, Berkeley.
    Includes bibliographical references (p.   ) and index.
    ISBN 0-292-79079-1 (cloth : alk. paper)
        1. Land use—Colorado—Planning.   2. Land use—Utah—Planning.
    3. Conservation of natural resources—Colorado.   4. Conservation of natural
    resources—Utah.   I. Title.
    HD211.C6W75   1993
    333.73'13'09788—dc20   93-7705

∧∧∧∧∧

Dad

^^^^^

CONTENTS

∧∧∧∧∧

PREFACE

Twenty years ago, I heard a Montana rancher use an expression that time has not weathered away. One muddy, cold April afternoon, he leaned against his manure-stained Ford pickup and told me a story. The man recalled how his neighbors to the west once rescued a flock of disoriented, nearly frozen snow geese during a blizzard. On that twenty-below-zero night, the family bundled up, went out into the gale, and herded the birds toward the wind-break of a barn where they fed bucket after bucket of grain and water to the exhausted creatures. The rancher said, "Those people stand from the ground." It was his highest compliment.

But what, exactly, is "standing from the ground"? In the modern ver-nacular, I suppose it means being "grounded," being fully alive and aware of where you are. But my friend was plainer than that. He meant that his neigh-bors knew their land and understood it. They belonged.

This book is a journey through Colorado and Utah in search of the places where people still stand from the ground . . . or are learning how to. I have tried to uncover the reasons why some communities take responsi-bility for conserving the beauty and ecological diversity lying at their door-steps and why others turn away. It has been both uplifting and frustrat-ing work.

We Americans are powerfully ambivalent about our home landscapes. Destruction and preservation exist in this culture like two interwoven strands of a rope. This has always puzzled me. I suppose it's no accident that I've spent my adult life wrestling with it.

∧∧∧∧∧

I have, at this writing, eighteen years' experience in conservation projects involving government agencies, corporations, The Nature Conservancy, and land trusts throughout the Rocky Mountain states. Five years of my life were spent as a planning director in Montana, and later I earned a Ph.D. in geography from Berkeley. I now teach geography and land-use planning at New Mexico State University and do conservation consulting work year-round. But my efforts to protect landscapes have not necessarily been driven by analytical, objective thought. I have helped secure the protection of over 100,000 acres of the West most of all because I am deeply indebted to the place.

In the spring of 1973, my life had taken a turn for the worse. Trauma was common in those turbulent days, and my share was sufficient to leave me confused, angry, and sick. After a hospitalization, surgery, and rounds of tests, I knew I was getting no better. The world was getting scary. I decided to make a go-for-broke move and headed out to Montana—a place I had never seen before.

I drove across the Great Plains alone in a red Opel with no radio. In the openness of that rolling country, it was clear what was happening. I was trying to find a way of life. It had come down to that.

The country began to change. First, there was the Little Bighorn. Then came the mirage-like crestline of the Beartooth Mountains. I drove past the Absarokas, the Crazies, the Bridgers, the Flint Creeks, the Sapphires, and on to Missoula at the base of the Bitterroot Range. After renting a place up Rattlesnake Creek, I resolved to try my best at graduate studies. But another kind of education was taking place each evening out on the trails above town.

Hiking was impossible at first. Months passed while I tried and failed to grow strong. I was at a loss. An offer to work as a land use planner appeared. I left school and moved to a county half the size of Connecticut with a population of 2,703. There were high country wilderness areas all around and creek after creek full of cutthroat, rainbow, brook, and brown trout. Ranchers invited me out for dinner. Their hay crops grew in flood-irrigated fields extending unbroken for 20 miles. Sandhill cranes nested in the lush valley grass in the spring. Elk wintered there when the snow grew thick in the mountains. This was a valley where the land was not a mute backdrop to life but the central character in the daily world of people. I began to heal.

It took a year or so, but one July morning I shouldered a backpack and used my own power once again to walk out on the land. I haven't stopped since.

Rocky Mountain Divide

Happiness and belonging bring health. Of this, I have no doubt. And land—beautiful, open land—is a medicine more powerful than all the pharmaceuticals likely to be found in the plants of the Amazon Basin. Ultimately, I wrote this book for one reason: to return a favor. It is an offering of thanks to the Western landscape and a carefully considered warning that this astonishing ecological and spiritual resource is now in tremendous jeopardy.

But judge for yourself. As you read these pages, take the time to get out your atlas and turn to the maps of Colorado and Utah. Follow along as we go from place to place. In the years to come the details of acres conserved and the names of land trust groups will change. The analysis of the underlying cultural and historical geography of the region may prove to be less perishable.

Some land-use planners may take offense at my critique of their profession. Two reminders are offered. First, I have nearly twenty years' experience as a planner. I am one of you. Second, it's an ill wind that blows no minds.

One last thing. If you become upset about the destruction of these landscapes, look around where you live and notice what is happening. It may be painful, but if it is, don't just fume and forget. There are pragmatic, effective actions that each of us can take to save the best of what is left of the homeland we love—wherever that may be. In the appendix, you'll find a list of land trusts and national conservation organizations. They can help.

In the end, it will be those acts of stewardship, not these words, which truly matter.

ACKNOWLEDGMENTS

This book arose from my Ph.D. dissertation in geography at the University of California at Berkeley. I wish to thank the Board of Regents for three years of fellowship support. Within the Geography Department my debt is profound. Dr. James J. Parsons showed heartening enthusiasm for this project and had confidence that I could be trusted to "just get out there and see" what I saw. I am proud to be a part of the "Berkeley School" of geography and the tradition of field-based cultural investigation begun by Carl Sauer. Special thanks also go to Ted Oberlander for giving freely of his knowledge of geomorphology and the Mojave. Thanks also go to Hilgard Sternberg, Barney Nietschmann, Bob Reed, Jay Vance, Orman Granger, Peirce Lewis, and Herbert Baker.

Geographer Tom Herman produced all the maps in this book with great skill and diligence.

Many fellow students, now geography professors, have expanded my world: Paul Starrs, Karl Zimmerer, Mark Blumler, Tom Eley, Scott Stine, Sally Horn, and Martin Lewis. The Geography Department office folks— Natalia, Charlie, and Doty—kept things moving along smoothly. Tim Sturgeon came through as a friend when it really counted. Bob and Pat Raburn showed extreme generosity in welcoming me into their home while I finished up my doctorate. Muchas gracias to Neusa Hidalgo Monroy Wohlgemuth for her goodness. A tip of the hat to all Malamuds.

I am eternally grateful to Chris Renarde for her wisdom, strength, and

∧∧∧∧∧

boundless faith. Lastly, a special note of appreciation goes to Dave Larson for his warm hospitality, table at Kip's, role as the griot of Berkeley lore, and encyclopedic knowledge of every aspect of life on the planet. Go Bears!

In Colorado, I would like to thank Marty Zeller, Don Walker, Hal Clark, the staff of The Nature Conservancy (TNC), and all the many planners and land trust members who spoke with me and from whom I learned so much. My love and respect go to Peggy Daugherty, McDuff, Amanda, Amos, and Annie. Irie.

In Utah, kudos to Cindy Smith for her thoughtfulness and generosity in lending me her house while I lived in Salt Lake. Appreciation goes out to Jay Edmunds for movies, humor, and "Hockey Night in Salt Lake." Dave Livermore of The Nature Conservancy was extremely helpful. Thanks also go to dozens of planners, agency personnel, and LDS Church members for their willingness to talk.

In New Mexico, my deepest appreciation goes to Leslie and Rutgers Barclay for an extended stay in their guest house. Gratitude also goes to Bill Waldman of TNC, Le Adams of the Trust Public Land (TPL), Tom Wolf, J.B. Jackson, Bill deBuys, Orlando Romero, Judy McGowan, Henry Carey, Barbara Jordan, Emilio Flores, Maria Varela, Michael Stewartt, and Mark Medoff. Sat nam to Swarn. Toni Delgado and Kelley Elkins showed me so much in the desert. Thanks also go to my compadres at the New Mexico State University Geography Department: Jerry Mueller, Al Peters, and Bob Czerniak.

In Montana, Jamie Kay, of Missoula's "Fine Lines," provided the most outstanding editing and creative help I could imagine. I am profoundly grateful to Chris, Faye, and Gretchen Field for superb barn housing, ice cream, and steady good cheer—I simply could not have done it without you. Much of what I know about land conservation has come from a long friendship with a gifted man, Bruce Bugbee. Thanks also go to Philip J. "Spike" O'Connell, John Harris, Bill Kittredge, Evan Denney, John Crowley, Bob Kiesling, Carla Burgess, Elmer Cyr, Merrill Riddick, Kathi Olson, all the folks in the Mineral County Courthouse, and the Pintlers. I give my love to Dorothy, John, Tommy, and Liz Whiston for treating me like family.

My best friend is Dave "Odeo" Odell—a kinte brother of the mountains and rivers for life.

And finally, to Mum and Cathy, the deepest love of all.

ROCKY
MOUNTAIN DIVIDE

^^^^^

PART I
LAND AND LIFE

∧∧∧∧∧

1

GEOGRAPHERS
AND LANDSCAPES

You can see a lot by looking.
—YOGI BERRA

This is a geography book, an exploration of conservation, land, and life in two of the Rocky Mountain West's most distinct cultural realms—secular, mythical Colorado and the mystical Mormon nation known as Utah. It is a search to explain how and why the wondrous, achingly beautiful landscapes of these neighboring states have grown so crowded and imperiled—and what is being done to save them. There are libraries full of books about the over-exploitation of resources within the West's vast expanses of public land. Stories about clearcutting of old-growth forests, rapacious mining enterprises, and misguided water development projects fill the region's daily newspapers. Given the national and often global significance of America's public lands, that attention is warranted. But what about the places where Westerners live, work, and raise their children? These cultural landscapes are of equal ecological importance, historical relevance, scenic beauty, and ethical meaning.

Privately owned lands keep the Rocky Mountain West in one piece. Most of the region's river corridors, visually prominent open space, prime ranchlands and farmlands, and historical/archaeological features are found in valleys lying far below the spectacular mountain crestlines which so often draw our attention. These private lands are also biologically critical for many wildlife species as winter range, birthing and nesting habitats, forage areas, and migration sanctuaries. Remote high country "islands" of public land, even if they weren't being assaulted by resource extraction industries

∧∧∧∧∧

and agencies, are simply insufficient to support viable regional ecosystems. Phrases now much in use like "Greater Yellowstone Ecosystem" bear witness to a growing realization: the boundaries which separate public from private land are not the same as the seasonal ranges of species, the sweep of scenic vistas, and the topography of watersheds. It is clear that political and ecological landscapes seldom correspond. Therefore, open, environmentally-sound private lands are as essential as national parks and wilderness areas, and they must be conserved if the Rocky Mountain West is to survive in any meaningful form.

However, today these very places are being destroyed. As California and other highly urbanized states grow increasingly polluted, violent, and unliveable, thousands of people are now moving to the Rockies in search of a rural refuge. In 1992, the four leading states in the growth of real estate sales were (in order) Utah, New Mexico, Colorado, and Montana. This threat is not a sudden, grandiose resource grab being fought by the Sierra Club or Wilderness Society. Few newspaper headlines are involved. No, the demise of the Western landscape is taking place because of the piecemeal subdivision and development of fields, floodplains, and forests . . . a few acres at a time.

This incremental wasting away of cherished local landscapes is the most vexing and least sensational of the West's environmental problems. The erosion goes on in near invisibility. There is a quiet Americanness about it all. The entire process of land development is advanced by city and county land-use planning agencies, elected officials, licensed realtors, banks, and solid, well-intentioned private citizens. There are laws, regulations, plans, surveys, permits, and codes. Given that all people are entitled to decent housing, it becomes difficult to find villains and assign blame. It is only when we draw back and look at our local landscape and remember what it was that the impact of the changes hits home. In that moment, the presence of what is absent starts to haunt us like the memory of a lost friend. It is then that the sinking hurt with no other name than mourning begins.

The destruction of ecological and open space resources is tragic enough, but assaults on "place" itself which fracture the union of land and culture have equally grave implications. Henry David Thoreau once implored: "Who are we. *Where* are we?" He was on to something. What will we be as individuals or as a society if we allow the degradation of our land's biological richness, agricultural productivity, and elemental beauty? If we are reflections of the places where we live, what will become of us if we exist in landscapes of fear, pollution, monotony, and lost promise?

Land and Life

The relationship between geography and a wonderful life is clear and compelling and rests at the very heart of this book. Although this is a story about the West, it illustrates a national dilemma. The losses mount from Bar Harbor to Big Sur. The central question in these pages is this: how can Americans conserve the best of what is left of our landscapes without trampling basic individual rights? This is something of a Zen Buddhist koan—a seemingly intractable riddle which has perplexed land-use planners and conservationists for decades. However, for every such conundrum, a "solution" can be perceived if we release our preconceptions and try to see the world plainly. And when grappling with the koan of "landscape saving," one is inevitably drawn into an ancient and beguiling way of seeing known as geography.

GEOGRAPHY

Geography is the exploration of landscapes: it is searching to find our way. Geographers seek out new landscapes to learn what they're like, how they work, and what they mean. From the land around us, we also come to understand a large measure of who we are. Our senses detect patterns and discover new features so that life can be kept both reassuringly familiar and excitingly fresh. All that we experience is retained in our memories. The shape of a local hillside, the sound of a street scene, the smell of childhood kitchens, and the indelible smile of our first love are hoarded as the treasures they truly are. We hold tight to our secret inner geographies like reassuring amulets which shield us from the uncertainties of a changing world.

A fondness for spatial memories is not found exclusively in malarial adventurers and musty arcane academics. A university degree or pith helmet is not required. It is enough to possess a compelling curiosity to strike off in search of whatever is out there. It may be North Dakota or the Tuamotu Archipelago—no matter. We may be looking for food, knowledge, or God— fine. For those of us with geographers' hearts, it is the search itself which moves us.

The Earth is teeming with geographers possessing an enlightened understanding of the landscapes around them. This knowledge has often come at a high price. In much of the world, if you make an errant spatial decision in search of food or new routes of travel, your final misstep will be remembered with reverence and humor around the campfires of your friends. What accumulates from generations of life-enhancing journeys and fatal pratfalls is a geographic learnedness in "uneducated" people which humbles bookish scholars and leaves their words sounding empty and untried.

The great navigators of Micronesia have long used stars, sea swells, winds, birds, and cloud patterns to guide them across thousands of miles of open ocean in small sailing vessels. These voyages are not only for trading and fishing, but provide honest pleasure and a real life. Australian Aborigines take ritual walks across waterless terrain to celebrate places sung into creation by their ancestors during Dreamtime. Mountain people in the Southern Appalachians can give detailed accounts of Civil War battles fought in the very forests where they stoop to collect ramps, fiddleheads, and mushrooms. When asked, these hill people can lead you unerringly through untrailed woods to mossy Confederate gravestones hidden by rhododendron, flame azalea, and kudzu.

People have known the Earth well long before Landsat images and computer-based Geographic Information Systems. Our survival has depended on it. However, while learning the lay of the land has sustained us physically, it has done much more. At times, our life trails have grown confusing and indistinct. While human beings are place-specific creatures, we are less forged by the landscape than perplexed by it. At times we are even in dialogue with the Earth.

The Beaver Indians of British Columbia and Alberta talk place names and revere maps. In *Maps and Dreams*, Hugh Brody recounts how hunting parties ponder their options of where to go:

> *From an accumulation of remarks comes something of a consensus. No, that is not really it: rather a sort of predirection, a combined sense of where we might go tomorrow. . . . People would then move out onto the land by a sense of rightness. . . . Some old-timers, men who become famous for their powers and skills, had been great dreamers. Hunters and dreamers. They located their prey in dreams, found their trails, and made dream kills, then whenever it seemed auspicious they could go out, find the trail, re-encounter the animal, and collect the kill. (36, 44)*

Old-timers sometimes became famous for the ornamented maps they fashioned from hides. The most esteemed of these elders were said to have made maps of the trails to heaven and were buried with these sacred guides.

Indians of the Fort Apache Reservation on Arizona's Mogollon Rim often speak the names of places encountered to friends after even the shortest trip. Tribal members say,

> *The land is always stalking people. The land makes us live right and looks after us. Our children are losing the land. It doesn't go to work on them anymore. They don't know the stories about what happened at these places. Some of us still know the names of places where everything happened. So we stay away from badness. (Basso, 95–96)*

Each portion of the Fort Apache Reservation is a repository for tribal, family, and personal history. Things happen "at the place where coarse-textured rocks lie above in a compact cluster" (Basso, 106). The practice is part mental mapping, part moral narrative. The landscape itself evokes stories of things prosaic and profound, tried and true, or tried and disastrous. It is a way of remembering, a kind of spatial folklore. The Earth reminds the Apaches of choices and consequences. They believe that despite poverty and racism, they are granted a measure of protection by being aware of the gifts of home land.

The Beaver Indians are Athapaskan people who remained north; the Apaches are Athapaskans who migrated in pre-Columbian time into what is now the American Southwest. It isn't surprising that some similarities show up. However, the ability to perceive great meaning in the landscape is not the bastion of one branch of "mystical Indians." It is clearly a worldwide trait, one that stands directly beside our baffling determination to destructively alter the Earth to suit our desires. The places we inhabit are unbiased chronicles of this internal struggle. Such landscapes are the ultimate cultural artifact, the complete vernacular archive of our travail, and can be studied for clues to the mystery of why human beings sometimes destroy and sometimes conserve the world around them. The investigation of this profound human paradox defines the geographer's craft.

Practicing this craft requires some unconventional traits. The basics are: a love of maps, a preference for small town newspapers, an ear for overheard conversations, attention to signs, a knowledge of plants, rocks, and rain, an eye for oddball juxtapositions, a delight in unexpected detours, and a pickup with gas enough to invite serendipity. Traditional research techniques sometimes help. But it is the *need* to travel in order to see, describe, and compare places that in the end "keys out" a geographer. Many of those holding this book fill the bill.

Geographers and Landscapes

The following bit of geography will focus on conservation, land, and life in the neighboring but antithetical states of Colorado and Utah. These places were chosen because they provide the West's most striking contrast of how cultures have settled, developed, and conserved private lands. Substantial differences were found in how land use is regulated and how environmental stewardship is expressed. The explanation of this pattern is far from simple. It is best advanced by going on the road for a look and following up on whatever seems important.

What results is a mulligan stew. This is a tale of land developers, land-use planners, The Nature Conservancy, local conservation groups known as land trusts, historical figures such as Brigham Young, Isabella Bird, and Buffalo Bill Cody, and modern characters such as John Denver, Hunter S. Thompson, and Robert Redford. Included are cautionary stories of shoddy land development running amok and uplifting examples of land-saving taking place against all odds. But mostly this book is about places—the valleys, towns, and cities which make up Colorado and Utah—and what is happening to them.

Some sections offer a straightforward analysis of information. Other passages are personal, opinionated, and anecdotal. The details of how landscapes look and feel have been relied on as much as leatherbound works of scholarship. No pretense of being a detached observer is offered. What follows is the work of a land-loving Westerner who has tried his best to be fair.

Before heading to the Rockies, some general understanding is needed of land-use planning as it is currently practiced in the United States. The contrast between the long-standing paradigm of land-use regulation and the rising movement for voluntary land conservation is striking. Across America, people are beginning to do something novel. They are quietly formalizing their personal responsibility to the Earth by taking pragmatic, legal steps toward leaving a rich legacy of open and unspoiled land. In most instances, the land-use regulation programs of city and county governments have not taken the lead. In fact, too often these agencies, established to shape tomorrow's landscapes, have either been unable or unwilling to help. It is a curious story.

2

LAND REGULATION AND CONSERVATION IN AMERICA

We have kiss'd away / Kingdoms and provinces.
—WILLIAM SHAKESPEARE, *ANTONY AND CLEOPATRA*

There are presently about 5,300,000,000 people in the world. Today, calling someone "one in a million" really means "There are 5,300 more just like you." During the past 40 years, our country has grown from 150 million to over 250 million people. In the early 1950s, a series of baby booms led to an unprecedented demand for new suburban housing lots, while a consumer-based, prosperous, postwar economy created a massive market for rural recreational tracts. This nationwide surge of land development caused open space, wildlife habitat, and agricultural land to be rapidly lost. Land speculation became like bowling—an American passion requiring no great skill to score well. As Levittowns spread across the countryside, more and more communities began to feel overrun and fearful of losing their way of life.

TRADITIONAL REGULATORY PLANNING

By the 1960s, population growth and land development pressures were so widespread that increasing numbers of communities which had never before "needed" land-use planning quietly formed commissions and hired staffs. This trend continued through the 1970s and early 1980s, aided by various programs from Federal agencies such as the Department of Housing and Urban Development and the Economic Development Administration.

The "Comprehensive Plan" lies at the heart of both long-standing and fledgling planning programs. These plans contain basic data on community

∧∧∧∧∧

demographics, land-use patterns, infrastructure, economic development, and social needs. They usually outline general goals and specific objectives designed to guide land development, protect the environment, and improve the quality of local citizens' lives. To enforce the comprehensive plan, most communities turn to subdivision regulations aimed at controlling the rate and type of lot splitting, to zoning codes for steering different types and densities of land use to preferred areas, and to floodplain regulations designed to keep structures out of wetlands and flood-prone terrain.

With the institutionalization of planning and the passage of regulations, citizens who were outraged by unrestricted growth felt that a significant corner had been turned. Some regulatory "victories" occurred, and there was a widespread belief things could only get better. To many, the rise of land-use planning was like the passage of the Clean Air and Clean Water acts and a natural continuation of the spirit of Earth Day. Their frustration would come later.

Government initiatives for conserving resources on *public* land have resulted in the establishment of national parks, wilderness areas, seashores, monuments, historic sites, and wildlife refuges. The legacy of John Muir, Gifford Pinchot, Theodore Roosevelt, and Aldo Leopold is tangible and widely enjoyed. Confrontational organizations and hard-headed individuals such as "Archdruid" David Brower also deserve our sincere thanks for keeping dams out of the Grand Canyon and for defending scores of nationally recognized natural features. However, at the local level, threatened *privately owned* lands possessing significant ecological or other value are unlikely to be protected by acts of Congress or lawsuits filed by national environmental-groups. These critical lands make up the cultural landscapes where we live. And here, aside from the creation of special case set-asides such as developed parks, the regulatory approach to controlling land use has a far different record.

Local government regulation of private land use is a rationally motivated but largely failed attempt at applied environmental ethics. Subdivision regulations in most parts of the country are treated like Prohibition Era edicts against having a drink. Zoning is often just wishing. Variances and re-zonings are routine facts of political life. Floodplain regulations rarely keep development out of low-lying areas but instead are geared toward managing Federal flood insurance programs and reviewing and approving engineers' plans for "flood-proof" structures. The traditional regulatory approach, by itself, has not significantly slowed the development of important portions of the private land base.

Land and Life

However, to conclude that land-use directives have "failed" is to misunderstand their real purpose. Land-use regulations focus on *how* land will be developed, not *if* it should be. The essential purpose of regulation is, therefore, not to conserve land but to see that developments are of the desired type and density, that design criteria are met, that local infrastructure can handle the increased load, and that effects on taxation and services are considered. These are rational economic, social, and technical concerns for communities. However, over the years, the whole subject of landscape conservation, which originally motivated so many people, often has been relegated to stock throwaway lines in comprehensive plans about protecting the local "quality of life," "scenic and ecological heritage," and "prime agricultural land." Many plans don't even accurately map the geographic distribution of lands whose conservation would further these goals.

Regulatory land-use planning makes people angry. Planners are routinely attacked by pro-development interests for being obstructionist and by conservationists for being ineffective. Over an otherwise calm four-year period as Planning Director in Mineral County, Montana, I twice received death threats and once had drunken gunfire erupt outside my cabin at 2:00 A.M. The fact that this is not unusual is both instructive and disconcerting. Planning is a difficult, underappreciated job where politics too often overrules reason as plans are drafted, debated, and discarded or passed, only to die quiet deaths on dusty bookshelves.

If you want to really rock planners back on their heels, ask them "what is good land-use planning?" It's a fair question. But for every planner who will take time to give you an honest opinion, there will be another whose response is a wisecrack, a blank stare, or a confession of uncertainty. Most planners enter the profession out of concern for the natural and social environment. Some days that's what they get to grapple with. However, most are at their wits' ends trying to stay just a few hours ahead of the paperwork. As a result, the big picture—known as long-range planning—doesn't receive its due.

To make matters worse, in many jurisdictions, public support for planning has waned as Federal community development and other grant programs shrink or disappear and local taxes are raised to make up the shortfall. As regulatory programs become more expensive, politicized, and highly visible, there is a surge of anger and doubt among citizens about what and who planning is really for. Planners and environmentalists often overreact by attributing the public's rancorous opposition to regulations solely to ignorance, greed, and backwardness. The public then retaliates by fighting, ignoring, or evading the planners any way they can. In the end, reasonable solutions often become lost in the shuffle.

It is abundantly clear that many private landowners make shortsighted, environmentally damaging land development decisions driven by a desire for profit. They react so strongly to planning programs because they see regulations—any regulations—as threats to their inflated sense of their land's value in the marketplace. Many depend on this mythical monetary value as their only real economic security for a rapidly advancing and uncertain old age. Landowners also tend to believe that God and the Constitution have bestowed on them the absolute right to use their land in any way they choose, even if such use would be obviously and significantly injurious to their neighbors. Thus, land-use regulation is widely seen as an anti-American, un-Godly "taking" of basic rights. This myth is so strongly held that no amount of Supreme Court case law in support of zoning and other forms of police power authority is likely to dislodge it from the psyche of American culture any time soon.

Attempts to force people to conserve land fall short for many reasons. Principal among these is the failure of officials to recognize and accept that conservation has more to do with socio-economics and psychology than planning theory and law. In *Losing Ground*, Erik Eckholm has written that planners and conservationists often don't fully grasp that:

> Land use patterns are an expression of deep political, economic, and cultural structures; they don't change overnight when an ecologist sounds the alarm that a county is losing its resource base.
>
> While socio-economic criteria may often exist in the minds of planners, there is an alarming failure to relate social and economic needs to acceptable forms of conservation. In the final analysis, if the small farmer and peasant [or U.S. landowner] cannot or will not take conservation to heart, no amount of model-building, empirical studies, environmental assessment, or legislation will result in the conservation of resources. (167)

Yet, when planners are confronted with an entirely predictable public objection to regulations, many behave like angry parents and prepare to discipline the unruly child. They regroup and, rather than question basic assumptions and planning methods, proceed to draft new sets of regulations with even less chance of passing or enforcing them. To many planners, the solution lies in finding the flawless code, the seamless paradigm, which can be overlain across the entire landscape and enshrined as gilded law.

The truth, then, is that traditional land-use regulations provide mostly the illusion of control. Ironically, they also ensure that battles will be fought over the same land again and again as real estate values rise and planning board goals change. Implementation of land-use plans through government regulatory authority often becomes a byzantine political struggle subject to delays, economically expedient zoning variances, and selective enforcement of rules. It can be an ugly business that by itself should not be relied on to conserve the special lands around us or to promote their "best" use. So, while legislated preservation of public land may sometimes work, widespread mandatory conservation of private land is about as possible as alchemy.

In recent years, a tremendous variety of government-initiated growth management systems have been devised to protect open space in states such as Florida, New Jersey, Vermont, Oregon, and California. Implementation techniques largely consist of large-lot agricultural zoning, property tax relief, right-to-farm laws, and in a few jurisdictions, the purchase or transference of development rights. Although growth management efforts have achieved sporadic success and are generally viewed as worthwhile, many landowners strongly oppose the specific land-use regulations and other control mechanisms required. In fact, growth management schemes have not been established in most jurisdictions due to political opposition, funding limitations, and the lack of staff training. As a result, development continues to spread across ecologically, scenically, and historically significant open land at a tremendous rate. The evidence lies before our eyes. In many cases, the houses and businesses built are important to the local economy but are poorly sited. The dilemma is this: If regulation is ineffective at conserving key portions of the landscape, what can be done?

VOLUNTARY CONSERVATION AND LAND TRUSTS

Americans concerned with environmental conservation have become deeply frustrated by the failure of government planning programs to protect cherished places from urbanization. Across the country people are reacting in two fundamentally different ways. Radical environmentalist "greens" and "deep ecology" advocates call for a retreat from technological advances and a return to a nostalgic arcadian way of life in rural areas. Greens apply confrontational political tactics and sometimes physical sabotage in attempts to halt growth and development.

A much more widespread and measured response comes from environmentally concerned moderate citizens. These individuals believe that in order to move forward in our efforts to conserve places, we must acknowledge a few things. First, governmental regulation of private property makes gun control appear uncontroversial. Second, nationwide this attitude is unlikely to change soon in ways which are meaningful for conservation. And last, if we want to do more than tidy up the infrastructure around a continuously developing landscape, we must meet with owners of key parcels and find ways to compensate them for *not* developing their land. This is not theory. Across the country, an ever-widening circle of conservation organizations is doing precisely that.

The rise of bioregionalism and the health of local historical societies emphasize that a compelling attachment to homeland still exists in the United States. Bioregionalism finds its expression in a potpourri of groups all over the country which range from crystal-clutching New Agers to conservative L. L. Bean-clad duck hunters. The Driftless Area Bioregional Network, Frisco Bay Mussel Group, Creosote Collective, and Northern Plains Resource Council are representative of these sometimes visionary groups who are dedicated to protecting the natural environments around them. Their members work by blunt or poetic words and by small and sometimes grand deeds to block the destruction of their chosen landscapes.

Local and regional land trusts are among the most practical of these organizations. Land trusts are private, nonprofit citizen groups which engage in land protection activities. Their mission is to conserve private lands of significant natural, scenic, and historic value. Land trusts do not function by rhetoric and earnest good intentions alone. Most trusts receive tax-exempt status from the Internal Revenue Service of the U.S. Treasury Department. This legal standing renders the value of gifts of land and conservation easements (development rights) made to the trust eligible as income and estate tax deductions for the donor. The conservation easement programs of land trusts and federal and state government agencies have permanently protected over 2.7 million acres of private land in the U.S.

THE HISTORY OF LAND TRUSTS

The recent renaissance of trusts disguises a surprisingly long history. Land trusts originated in Massachusetts in 1891 with the formation of the Trustees of Reservations. This Boston-based group arose in response to land development spurred by a growth in U.S. population from 38 to 76 million between 1870 and 1900. By 1950, there were 53 trusts in the country, mostly in New England. By 1965, this had increased to 132 in 26 states, nearly all located in the Northeast and the Middle Atlantic regions. At that time, none existed in the Rocky Mountain West.

The environmental movement of the 1960s rekindled America's conservation traditions. Much attention was focused on problems such as air and water pollution, national park management, and the formation of wilderness areas. However, land trust activity also gained momentum when a few conservationists took a deep breath and acknowledged that the sacred Earth was also real estate.

Conservationists, who previously could appeal to emotion or to a higher court to protect ecosystems, watched as new opportunities formed in front of them. Conservation easements, land exchanges, partial development designs, and an array of other voluntary, negotiated, compensating land protection techniques began to be widely used. Talking real estate was at last coupled with "speaking the names" on the land. This breakthrough, combined with a widening frustration with the ineffectiveness of regulatory land-use planning, produced new energy. By 1975, the number of land trusts had more than doubled to 308. New England and the Middle Atlantic states remained the bastion, but increasing numbers of trusts were formed in the Midwest, South, and Pacific Coast regions. The Rocky Mountain states had 3 trusts, all in Colorado.

In 1981, the land trust "movement," then totaling 431 groups, gained its most important ally, the Land Trust Alliance, then known as the Land Trust Exchange. The central purpose of this organization is to facilitate communication between land trusts regarding techniques, experiences, and ideas so that more land will be protected. The organization is a clearinghouse of information and the switchboard for land-saving networks. It now also has a lobbyist in Washington, D.C., to look after land trust interests.

Since 1981, the number of land trusts has exploded. There were 535 trusts in 1985. As of 1992, 889 land trusts existed in 48 states, the Virgin Islands, and Puerto Rico (Land Trust Alliance) (Map 1). Only Arkansas and Oklahoma have no trusts. This total of 889 groups includes some quasi-governmental organizations with conservation easement programs. Trusts have formed at the rate of more than one each week for the last five years. Land trust membership now stands at 775,000 people, up 300,000 since 1984.

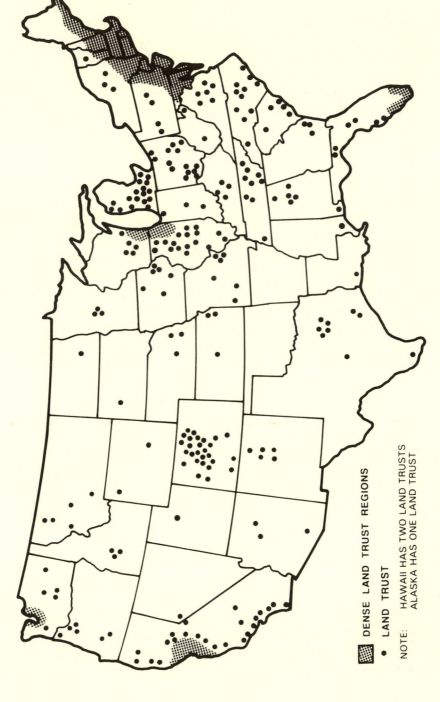

Map 1. United States Land Trusts, 1992 (Source: Land Trust Alliance, 1991 and 1992)

DENSE LAND TRUST REGIONS

• LAND TRUST

NOTE: HAWAII HAS TWO LAND TRUSTS
ALASKA HAS ONE LAND TRUST

LAND TRUSTS TODAY

One would search in vain to find comparable growth in support of any other grassroots movement. Through the use of various innovative tools, America's land trusts have protected nearly 3 million acres from development. In the last two years alone, trusts have kept more than 630,000 prized acres from being subdivided and built on. However, the distribution of trusts is still predominantly in the eastern United States. New England has nearly 400 trusts, 42 percent of the national total. Massachusetts and Connecticut are home to more than 100 trusts apiece. Maine has over 65. The mid-Atlantic states have 150 land trusts, most of these in Pennsylvania and New York. The Far West has 116 (77 in California), and the Great Lakes region has 98 trusts. The Rocky Mountain states of Colorado (27), New Mexico (5), Montana (5), Idaho (3), Wyoming (2), and Utah (1) have a total of 43 land trusts.

Nationwide, there has been a tremendous diversification of the "land trust" idea. Their names bespeak their goals and geographic settings: Big Sur Land Trust, Humboldt North Coast Land Trust, California Coastal Conservancy, Marin Agricultural Land Trust, Sonoma Land Trust, Kings County Farmlands Program, Colorado Open Lands, the Aspen Valley Land Trust, Montana Land Reliance, Jackson Hole Land Trust, Texas Conservation Foundation, Platte River Whooping Crane Habitat Trust, Ozark Regional Trust, Brandywine Conservancy, French and Pickering Creeks Conservancy, Natural Lands Trust, Hillside Trust, Adirondack Land Trust, Scenic Hudson Inc., Block Island Land Trust, Vineyard Open Land Foundation, Barnstable Conservation Commission, Society for the Protection of New Hampshire's Forests, Ottaquechee Land Trust, Lake Champlain Islands Land Trust, Maine Coast Heritage Trust, and on and on they go. Community land trusts, an offshoot of the movement, help people in cities and rural hamlets carry out self-help housing, community gardening, park development, and other programs designed to address poverty, crime, and the loss of neighborhoods.

Over 75 percent of all trusts are geared toward protection of ecological resources and open space. More than half of all trusts are involved with protecting recreation areas, wetlands, or watersheds, and about half work to save river corridors, lakes, forests, or farmlands. Greenways and historic sites are on the agenda for about one-third of all trusts.

The Land Trust Alliance conducted a national survey in 1991 which revealed that the success of trusts is extremely uneven. (Land Trust Alliance) Some 41 percent of the respondents had protected under 100 acres; an additional 37 percent had conserved less than 1,000 acres. While acreage alone is a crude measure of "success," given the primary goal of most trusts to "save landscapes," acreage protected provides at least a rough measure of effectiveness. Trusts often struggle because, as of 1991, most had entirely voluntary (54 percent) or part-time (16 percent) staffs. Only trusts with full-time staffing (30 percent) were likely to have had formal training in the implementation of land-use planning and conservation real estate projects.

The 1991 survey also revealed that trusts are severely underfunded: 54 percent had yearly budgets of less than $10,000. While one-quarter of all trusts had budgets over $100,000, about a third of these focused on environmental education, ecological research, historic preservation, alternative agriculture, and the management of existing nature reserves and botanical gardens. Only 22 of the 124 trusts with annual operating budgets of $100,000 or more had conserved more than 10,000 acres. However, this small core of highly successful groups tallied some of the highest acreage totals.

California's Marin Agricultural Land Trust has used conservation easements (donated development rights) to protect 21,000 acres of dairy farms, ranchlands, and scenic coastal property just north of the rapidly expanding San Francisco metropolitan region. Pennsylvania's Brandywine Conservancy holds conservation easements on 18,000 acres of farmland, forests, and wetlands within the Philadelphia/Wilmington urban corridor. The Maine Coast Heritage Trust has conserved 28,000 acres of coastal agricultural lands and islands, much of it near Acadia National Park. The Maryland Environmental Trust has secured conservation easements on 26,000 acres of coastal farmland. The Virginia Outdoors Foundation manages 60,000 acres of conservation easement open space within greenways and river corridors. The Society for the Protection of New Hampshire's Forests is the country's most accomplished land trust, with over 850,000 acres conserved.

Government can also play a highly visible role. Some jurisdictions are even taking the lead when land trusts prove ineffective. In 1988, California voters passed Proposition 70, the so-called "California Wildlife, Coastal and Park Land Conservation Act." With this proposition, citizens voted to finance a $776 million bond issue to be used by counties to purchase development rights (conservation easements) to protect coastal habitat, scenic open space, and farmland. In 1990, Floridians funded a $3.2 billion bond. To date, government Purchase of Development Rights (PDR) programs have been used by nine states to protect 205,000 acres at a cost of $400 million. In addition, the establishment of Adirondack Park in upstate New York and the innovative New Jersey Pinelands Preserve are other notable success stories arising from government initiative.

THE NATURE CONSERVANCY

Most local and regional land trusts were preceded by the well-known, internationally active trust, The Nature Conservancy (TNC). The Nature Conservancy's mission is to maintain the Earth's biological diversity by saving threatened and endangered species from extinction. To date, TNC has conserved some 4 million acres in the U.S. (Land trusts may be conceived of as mini-Conservancies with their own agendas.)

TNC has assembled the largest private preserve system in the world. Its 1100 preserves protect over 7 million acres across the globe. The 400,000 members of TNC annually donate about $40 million, or half of the organization's income. The rest comes from big donors, project income, and sales of donated lands without significant ecological value.

TNC is the heavyweight champion of fundraising largely because it is determined to stay out of political advocacy and remain project-oriented. In 1986, for example, TNC raised $73.4 million in cash and received more than $52 million in donated property. Forty million dollars of this was then used to buy or negotiate conservation easements on property valued at $166.6 million. The remainder of this income was put into an $80 million revolving project fund, long-term programs, and endowments. As a result of this kind of financial astuteness, TNC has assets in excess of $400 million.

Two recent projects illustrate The Nature Conservancy's effectiveness. In 1989, TNC completed the largest private conservation transaction in history with the acquisition of the 321,000-acre Gray Ranch in New Mexico for $19 million. A year later, media magnate Ted Turner donated two conservation easements to TNC which spared 130,000 acres of key Montana habitats from development—some 10 percent of the private land in the Greater Yellowstone Ecosystem. Local and regional land trusts generally operate on relatively tiny budgets and have not approached the record of The Nature Conservancy.

LAND TRUSTING

The Nature Conservancy has been a leader by example and has been involved in cooperative projects with land trusts and government agencies for years.

The Land Trust Alliance continues its work of spreading the land trust message across the country. The Trust for Public Land (TPL) and the American Farmland Trust (AFT) also provide small trusts with opportunities to learn the ropes. TPL has an ongoing program to train and provide technical assistance to local land trusts. Organizations such as the Conservation Foundation, Rocky Mountain Elk Foundation, the National Trust for Historic Preservation, the Sempervirens Fund, the Conservation Fund, and many others fill an increasingly diverse, professional, and occasionally crowded playbill of what former Secretary of Interior Stuart Udall has called "earthkeepers." This network of conservationists is continually pioneering new methods of protecting unique landscapes, rare habitats, and historic sites and structures. Cooperation is the general rule, and productive alliances are often formed during the creation and implementation of complex projects.

Across the country, there is a growing realization that we as individuals must confront our responsibility for the land. Government will help us if we wish it to, but the choice is ours. Land trust membership cuts across all usual barriers of region, race, politics, wealth, and gender to bring about voluntary stewardship of our natural and cultural inheritance. Land trust members are practitioners of an applied folk geography which is powerfully resonant with our traditions of community democracy.

Today, when land trusts practice their special craft, it is with increasing confidence, skill, and creativity. Staff members are now as likely to have taken courses in estate tax planning and fundraising as Botany 101. To be successful negotiators, they must be equal parts planner, lawyer, poet, and psychologist. They appear to have taken to heart science fiction writer Robert Heinlein's belief that a person should be able to build a house, write an essay, make an omelet, and track a deer—specialization is for insects. Land trust members are motivated by an emotional attachment to where they live. However, they move forward with the clear knowledge that it is the synthesis of values and legal techniques which makes land-saving happen.

CONSERVATION EASEMENTS AS LAND SAVING DEVICES

The single most important tool of a growing assemblage of voluntary, negotiated land protection techniques is the conservation easement. A conservation easement is a less than total interest in a parcel of land which is conveyed by a landowner to a government agency or IRS-qualified tax-exempt conservation organization (such as a land trust). An easement is essentially the "development rights" portion of land ownership. The purpose of such easements is to protect significant recreational, ecological, agricultural, open space, and/or historic values of the land. Easements assure that certain rights of use are prohibited or controlled.

Conservation easements are not a government regulatory tool. Landowners voluntarily donate or sell them. Restrictions are negotiated between the landowner and the group receiving the easement. These restrictions are based on thorough ecological analysis of the land. What results is a formalized, jointly held vision of how a property will be cared for in perpetuity. Easements are less commonly granted for a specified term of years.

Granting a conservation easement results in a legal division of ownership. The conservation organization holds rights of land use which, if exercised, would damage or eliminate the property's significant conservation resources. The landowner continues to own the land and use it within the limits of the easement. The easement holder must monitor the land each year to see that the terms of the agreement have not been violated. Violations which cannot be corrected voluntarily are solved in court. This rarely occurs.

Each conservation easement is unique and requires its own clearly stated, specific restrictions. Land uses such as subdivision and development, clear-cutting timber, and overgrazing are commonly prohibited. Easements run with the land: all subsequent owners of a parcel under protection must comply with the restrictions. There is nothing more durable in American real estate law than title, even if it is partial title. This, and the landowner's commitment to stewardship, are what give conservation easements their strength. Unlike police power measures, which focus on regulating *land use*, easements also reconfigure types of *land tenure*.

The voluntary granting of a perpetual conservation easement to a qualified receiver is a charitable conveyance under Federal law and IRS regulations. For income tax purposes, the transaction is similar to giving a cash donation to a church or the United Way. Therefore, the public pays for conservation easements via forgone tax revenues. The amount of income tax and other tax benefits generated by an easement is determined by comparing the appraised value of the land before and after land use rights are donated. The difference equals the value of the gift. The landowner continues to hold title and to control public access and must pay taxes on the land at values related to the ownership rights retained. He or she may freely sell the property on the open market at whatever price it brings.

Conservation easements have been used in this country for over 90 years. Their initial applications were principally as scenic easements purchased by government agencies such as the National Park Service for the Blue Ridge and Natchez Trace parkways, the Wisconsin Highway Department for the Great River Road, the U.S. Forest Service for the Sawtooth National Recreation Area, and the U.S. Fish and Wildlife Service for waterfowl habitats. Today, local, regional, and national land trusts have protected over 1 million acres, and government agencies over one-half million acres, using *donated* conservation easements. The Montana Land Reliance leads the nation's local land trusts in the application of donated easements with 96,000 acres conserved.

Conservation easements are only one component of a complex collection of voluntary techniques employed by land trusts and government agencies. Approaches to protecting land can be divided into:

1) Status quo: continuation of historic land use and stewardship by private landowners

2) Negotiated term agreements for conservation and public access

3) Conservation easements: acquisition of limited interests in real property by appropriate agencies or organizations

4) Outright ownership: full acquisition of land parcels by government or land trusts

The last and first items in this list are one-sided. Either the private landowner or the public assumes the entire responsibility for conservation. The other two choices, term agreements and conservation easements, involve a partnership between private landowners and the public. Some form of compensation is involved, be it personal, financial, and/or spiritual. As Norman Myers writes in *The Sinking Ark*, "You can persuade some to protect land, but ultimately we must recognize the sacrifice and compensate it" (233). It is compensation of some sort which makes conservation take place.

REGULATIONS VS. VOLUNTARY CONSERVATION

There are many fundamental differences between voluntary land conservation and traditional systems of land-use regulation. The most compelling is that the purpose of the voluntary approach is to protect important lands by dealing with land ownership; the purpose of regulation is most often to meet more vague public goals through the application of land-use laws. Despite the inherent ecological variety within a political jurisdiction, regulatory systems sometimes ignore environmental differences because of statutory mandates against "arbitrary and capricious" land-use decisions. While reasoned findings of fact can be used to zone or otherwise block development of sensitive lands, this is not always done due to the possibility of legal challenges. The land trust approach is specifically designed to differentiate land based on its visual uniqueness and ecological richness.

TABLE I

Voluntary Approach	Regulatory Approach
Purpose: to conserve land and encourage sustainable development	Purpose: to assure that development meets design specifications; to protect public health, safety, and welfare
Permanent: perpetual legal interest acquired	Temporary: laws, regulations, politics change
Specific common ground of agreement identified	Government decision in the general "public interest"
Compensating: meets landowners' economic and nonmonetary needs	Noncompensating: limits rights
Rapid implementation	Slow implementation; need to regulate actions on same land often
Politically acceptable	Politically controversial
Flexible: designed to fit land and landowner	Rigid: uniform rules for all lands and landowners
Many issues considered	Single issues predominate
Occurs prior to development decision	Occurs after development decision
Widespread application: useful over entire landscape	Limited application: only certain lands and land uses are regulated
Addresses full range of land-use practices	Addresses only certain land-use practices
Encourages sense of individual responsibility to land and community	Provokes defensiveness and divisive confrontations
Fosters independence	Fosters dependence
Allows creativity	Limits creativity
Vernacular	Regimented

HYBRID APPROACHES

The voluntary approach is not a panacea. It can, however, be an effective complement to traditional planning. Land trusts often protect key portions of landscape which hold a place together, while planners enforce regulations ensuring liveable and cost-effective development of other lands.

Innovative park, open space, and greenways programs exist all over the country. Numerous excellent park and open space systems have been assembled by governments through land purchases which predate the rise of land trusts. In other cases, land trusts buy or acquire an option to buy an important parcel. They then hold it until the community either raises funds to purchase the property or decides through its planning program what level of development is compatible with long-term goals. Watershed protection, recreational corridors, and farmland preservation are areas where trusts and regulatory agencies have worked together successfully. Partial development designs (with conservation easements) also are being used now to implement government Planned Unit Development (PUD) zoning districts. In such cases, houses are clustered on the least sensitive lands with the surrounding open space preserved by an easement. A few jurisdictions have attempted "Transfer of Development Rights" (TDR) programs. These systems designate conservation areas from which development rights can be sold and "sent" to areas where dense development is encouraged. By 1992, some 40 communities were attempting TDRs. However, in most cases, the public's confusion over the complexity of the technique has greatly limited its effectiveness. While 36,000 acres in the United States have been protected using TDRs, over 24,000 of these acres were set aside in one jurisdiction—Montgomery County, Maryland.

CONCLUSION

Far too many city and county land-use planning professionals remain either unaware of or strangely resistant to the new alternatives advocated by land trusts and forward-thinking planners. Some local planning staffs even seem threatened by the new methods and attempt to minimize their practicality. This isn't surprising. Voluntary techniques of land-use control are noticeably absent from the curricula of university planning programs. The editors of professional planning journals also largely ignore the shifting social currents now taking place. Therefore, when planners react defensively to the innovations practiced by trusts, it is often because they haven't got a clue what's going on or the time to find out. Fortunately, this state of affairs is changing as more and more jurisdictions shoulder the task of land-saving by establishing open space protection programs, which function much like land trusts. Yet, it is a bitter irony that in far too many cases the current "movement" for land protection has not been led by those formally charged with the responsibility to create liveable landscapes, but by the public the planners have fought so long to regulate. Many land trusts are also staffed by former land-use planners who quit the profession in frustration. One prominent land trust director even describes herself, half in jest, as a "recovering planner."

In the excitement over the nationwide expansion of land trusts and the creation of new hybrid planning systems, it is tempting to assume that every state will soon wholeheartedly embrace the notion of private land conservation. But this would be geographic nonsense. The trajectory of the land trust "revolution" is strikingly different from place to place. In some states land trusts are absent or doing poorly. In others, some land trust advocates are now concerned that too many groups exist and that competition for projects is becoming counterproductive. This frustrates and puzzles conservationists, and many are now asking why such disparities exist. Part of the "why" is because of the distinctive historical and cultural geographies of different regions and because of the fundamentally dissimilar interior geographies of the people who live there. The approaches used by land trusts and innovative planning and open space agencies are not universally transferable to every setting, but few have waded far enough into the geography of places to offer an explanation.

A close-up look at a culturally diverse region—the Rocky Mountain West—may help clarify the circumstances in which land conservation and enlightened land-use planning achieve acceptance or are dismissed. This exploration is particularly critical in the Rockies because there is so much at risk on the ecologically rich private lands of this region mostly known for its national parks and wilderness areas. But there is something even greater in jeopardy. The Rocky Mountain West itself, a chimera long extolled as an endless wave of unspoiled land, is being transformed from open landscapes with scattered communities to congested swaths of urban sameness and splatters of recreational development. This ironic reconfiguration, occurring with increasing speed, has profoundly important historical roots.

3

THE ROCKY MOUNTAIN WEST: SEARCHING COUNTRY

A man arrives as far
as he can but not as far as he wishes.
—VASCO DE BALBOA, LETTER TO
KING FERDINAND OF SPAIN, JANUARY 20, 1513

The Rocky Mountain West is searching country (Map 2). We have always arrived here empowered by myths, fear, and boundless hope. It is a land so different than where we come from, so grand and preposterous, we often cannot see it for what it is. It is a kind of *tabula rasa* landscape—a place where if you squint into the setting sun, the land seems transformed into a more familiar and compliant reality—but the illusion cannot be sustained.

Since the Pleistocene Epoch, when paleo-Indians ventured here from Asia, human beings have staked conflicting, overlapping claims to this awesome expanse. Yet, a claim is only something professed; it expires when you cannot sustain your dream or another's declaration is more powerfully defended.

With the settlement of New Mexico, Eastern America, and the West Coast, tribal dislocations moved like shock waves across the continent. The uneven spatial and temporal introduction of horses and rifles made certain tribes more mobile and powerful and better able to seek out and control new territory. Over the centuries, Indians, fur trappers, explorers, priests, miners, railroaders, homesteaders, loggers, the Defense Department, land developers, and conservationists have all jostled for position. As a result, claims and counterclaims reverberate throughout the West like countless sets of ripples in a rainswept pond.

The Rocky Mountain West is a realm where the land's the thing, and simple movement is believed to be magic enough to conjure up a more ful-

∧∧∧∧∧

filling future. Historian Frederick Jackson Turner often exaggerated the role of The Frontier in shaping social systems, but he was justified in stressing the importance of "This perennial rebirth, this fluidity of American life, this expansion westward with its new opportunities" (Turner, 2).

However, a lot more is going on in the West than economic imperatives, and the settlement of the region cannot be reduced to the march of expanding capitalism alone. For hidden behind settlers' dreams of wealth has often been a nagging sense of something else. Throughout history, certain people who land in the West and "stick it out" discover the region's native paradox: it is a province where settlement precedes the true search. This quest is initially driven by a yearning to explore and freely exploit a spectacular open land. As Wallace Stegner noted, "Geography, at least, is one matter in which the Westerner can excel and in which he takes pride" (Stegner in Guthrie, viii).

Yet, when the scenery begins to wear off, Westerners still find themselves looking at outcrops and listening to the cold November wind blow through the trees. What are they after? Gold? Board feet of timber? Most likely so. But beyond these economic rationales, often buried by politics, embarrassment, or a lack of words, lies a deeper truth. The search of many Westerners is not for money or the meaning of life, but for the experience of life—for the McCoy.

DREAMS AND HARDWARE

The settlement and exploration of the West was made possible by the timely coupling of dreams and hardware. Land-claiming was not initiated by agrarian folk struggling to tame the wilderness with their bare hands. It was a complex social phenomenon promoted by Federal land offices, railroads, and other gifted hucksters and implemented by the foundries of the Industrial Revolution. Molten metal was fashioned into instruments to rid the land of Indians and to make settlement at least plausible. No righteous agrarian dream supported humid-zone settlers in unknown arid and semiarid terrain: it was the railroad, barbed wire, repeating rifles, steel plows, windmills, telegraphy, and tin cans of Georgia beans. Much later came four-wheel drives, snowplows, powerlines, trailer houses, propane tanks, recreational skis, computerized land survey gear, satellite dishes, and nuclear bombs. Industry led to the settling of the West and is responsible for its cycles of economic growth.

The Rocky Mountain West

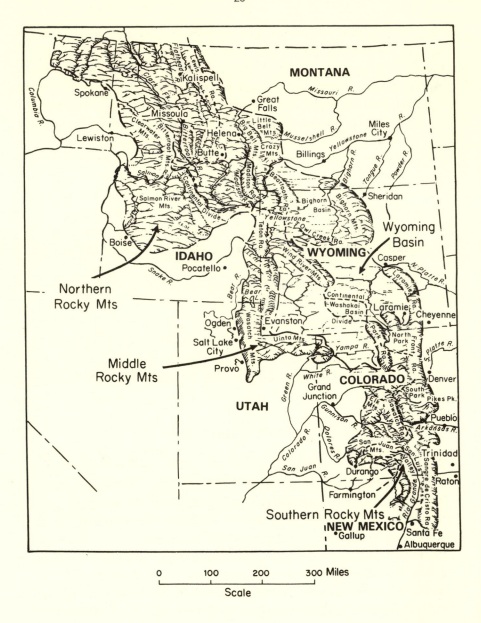

MONTANA

Columbia R.

Spokane

Lewiston

Kalispell

Missoula

Helena

Butte

Great Falls

Billings

Missouri R.

Musselshell R.

Yellowstone R.

Miles City

Bighorn R.

Tongue R.

Powder R.

Sheridan

IDAHO

Boise

Pocatello

Salmon River Mts.

Snake R.

Northern Rocky Mts

WYOMING

Bighorn Basin

Wind River Mts.

Owl Creek Ra.

Continental Washakai Basin Divide

Wyoming Basin

Casper

N. Platte R.

Laramie Ra.

Laramie

Cheyenne

Middle Rocky Mts

Ogden

Salt Lake City

Provo

Evanston

Bear R.

Wasatch Mts

Uinta Mts.

Yampa R.

Park Ra.

North Park

Front Ra.

S. Platte R.

Denver

UTAH

Green R.

White R.

Grand Junction

Gunnison R.

Colorado R.

Dolores R.

San Juan R.

COLORADO

South Park

Elk Mts.

Sawatch Ra.

Pikes Pk.

Pueblo

Arkansas R.

San Juan Mts.

Durango

Farmington

San Luis Valley

Sangre de Cristo Ra.

Rio Grande

Trinidad

Raton

Southern Rocky Mts

NEW MEXICO

Gallup

Santa Fe

Albuquerque

| 0 | 100 | 200 | 300 Miles |

Scale

Map 2. The Rocky Mountain West (redrawn from Hunt, fig. 13.1, p. 246)

Land and Life

The process of settlement has been continuous. Only in the tidy world of Frederick Jackson Turner could the precise date of 1890 be given for the end of "The Frontier." In *The Legacy of Conquest: The Unbroken Past of the American West*, Patricia Limerick chronicles Western history's hamstringing by the widespread belief that a discontinuity exists between the frontier and modern life. This fallacy continues to leave us adrift without the historical and geographic bearings required to understand the region's current environmental, economic, and social dilemmas. Limerick believes that once history and geography are:

> *reconceived as a running story, a fragmented and discontinuous past becomes whole again. With its continuity restored, Western American history carries considerable significance for American history as a whole. Conquest forms the historical bedrock of the whole nation, and the American West is a preeminent case study in conquest and its consequences. (27–28)*

And conquest is always driven by the powerful motivating beliefs of prescient leaders.

THE GREAT DIVIDE

The right to acquire and re-sell real estate without significant interference is a truth long held to be self-evident in America. George Washington was a prominent land speculator during the settlement of the Allegheny Mountains following the Revolutionary War. Later Thomas Jefferson and Michel Crevecoeur stepped forward as stirring proponents of the yeoman ideal and the moral superiority of agrarian life. This notion has a history which spans from Aristotle to Wendell Berry. In eighteenth- and nineteenth-century America, the image of the yeoman and his "freehold" possessed enormous power as a force shaping land transactions and historical geography. The survey ordinances of 1785, 1787, and 1796 were designed to create rectangular order upon a continent of which little was known. These Federal bills were ambitious attempts to transform the land from chaos to cosmos, but it would be decades before surveyors reached the Rockies.

During the early 1800s, Meriwether Lewis and William Clark, John Fremont, James Bridger, Stephen Long, and many others provided geographic information on the new lands of the Louisiana Purchase and various

western cessions. However, their accounts often fueled unrealistic perceptions of promise or evoked an endless Great American Desert. Despite, or
perhaps because of, a paucity of reliable information on climate, soils, and
water availability, Americans began their infatuation with "The West."

The U.S. General Land Office was formed in 1812 to facilitate the west-
ering impulse sweeping the country. However, word of mouth—repeating
promoters' lies—was the real motivating force of the movement. In *Virgin
Land*, Henry Nash Smith wrote,

> *The image of a vast and constantly growing agrarian society in the
> interior of the continent became one of the dominant symbols of the 19th
> century American society—a collective representation, a poetic idea that
> defined the promise of American life. (123)*

A battery of Federal legislation was passed during the succeeding decades to
enact the manifest wisdom of settling a "vacant" land with righteous farming
folk. "Western" settlement in this period meant Ohio, Illinois, and our
present Midwest. By the 1830s, Federal land subdivision sales accounted for
40 percent of national revenues, and budget surpluses became common.
Speculative fever rose until 1837, when the unsupported paper money of
scores of grossly overextended local banks collapsed and took land prices
with them. A national economic depression ensued.

The cadastral partitioning of the Rocky Mountain West itself was initiated in the seventeenth century by the King of Spain, who made land grants
in New Mexico to favored soldiers and friends. However, for most of the
American West, subdivision began during the administration of the General
Pre-emption Act of 1841. As settlement outpaced land surveying, squatters
were allowed to buy a 160-acre "entry" for $1.25 an acre. They could "pre-
empt" others by purchasing land without competition. The General Pre-
emption Act, like its earlier versions, was much abused. Fraudulent filings
rendered the process an administrative quagmire.

In 1860, following the Mexican cession, Gadsden Purchase, and other
acquisitions, the Republican Party attempted to gain national political advantage by pledging to enact a sweeping law allowing free homesteading
throughout the unsettled portions of a newly immense America. The Free
Soil Party (Democrats) and *New York Tribune* publisher Horace Greeley's
National Reform Association also lobbied hard for the provision of free land
for settlers far from the grasp of speculators. An emotional Greeley wrote:

Wherever settlement has begun land speculation has already been very busy. . . . The settler must either go out of the world or pay the speculator's price, which is often three, five, and even ten times the Government's price. This speculation is perpetually operating to scatter, to retard and barbarize our pioneer settlements. . . . Banish the speculator or break up his pestilent calling. (in Hibbard, 361)

In 1862, Congress passed the Homestead Act, which offered settlers 160 acres free of charge with title passing to them after five years of productive occupancy. The vision was of an agricultural fee-simple empire. The Homestead Act was a piece of wartime socioeconomic and geopolitical legislation with many aims. However, with the cavalry called home to fight the Civil War, it was mostly hoped that Northern settlers would form a barrier against Confederate advantage and Indian uprisings.

In the early years of the Homestead Act, a significant share of the land chosen and patented by settlers remained in small farms. However, as the Rocky Mountains began to be intensively claimed, the process changed. Between 1880 and 1904, 22 million of 96 million acres of Homestead entries were quietly bought by speculators for $1.25 per acre under a loophole known as the Commutation Clause. A homesteader could call it quits after six months from the date of filing on a property. A frontier realtor would then pay the bill and acquire title. Resale values rose as large monopolistic holdings were accumulated in many areas through commutation transactions.

Congress attempted refinements, but cagey entrepreneurs always seemed to be a step ahead. Despite fraud and the foolishness of such small homestead tracts, settlement continued apace. Washington was pleased. After all, the main point was to fill the West with people. It was believed that this would assure an Edenic social stability and add demographic credence to the bravely shouted but audacious cry that America was a truly continental country. There were 214 million acres claimed under the Homestead Act between 1862 and 1923. Although never meeting all the lofty expectations which surrounded its passage, the Homestead Act brought many people west and some of them actually stayed.

In the West, weather has often been mistaken for climate—and vice versa. Beyond the hundredth meridian, wet spells have long been celebrated as a return to "normal" climatic conditions, with the inevitable, cyclic droughts portrayed as aberrations of weather or as unforeseen disasters. A series of years with ample moisture for crops and other needs has, since the beginning of settlement, rendered otherwise down-to-earth people memory-less and prone to self-deception. This is because aridity has been widely seen as "uncivilized" and damned inconvenient. People tried to wish the dryness away. In his 1844 book, *Commerce of the Prairies*, Josiah Gregg voiced the sanguine view that would shape the era:

> *Why may we not suppose that the general influences of civiliza-*
> *tion—that extensive cultivation of the earth—might contribute to the*
> *multiplication of showers? (193)*

The influential Horace Greeley and his agricultural editor, Nathan Meeker, heartily supported this idea, as did a nation itching to expand. As a result, "rain follows the plow" was invoked like a prayer across the West. However, for many Homestead Act settlers, the only moisture to fall on dusty furrows was their own sweat.

If the Christian yeoman's holy plow couldn't civilize the climate, then perhaps the Pagan's sacred tree could. The need to believe the West could be made humid ran deep and beyond reason. It was easy to dismiss doubt since even government experts and the writers of impressive-looking books said the dream was true. Between 1867 and 1879, Ferdinand Hayden, Clarence King, Lieutenant George Wheeler, and Major John Wesley Powell carried out the great geographical surveys of the West. These men were the first to methodically assess the region's potential for agriculture, irrigation, grazing, minerals, timber production, and settlement. However, not all their findings have stood the test of time. Hayden was a highly respected figure who chose to perpetuate folklore rather than challenge it:

> *It is believed that the planting of ten to fifteen acres of forest trees on*
> *each quarter section (160 acres) will have a most important effect on cli-*
> *mate, equalizing and increasing the moisture and adding to the fer-*
> *tility of the soil. (Hayden, 135–136)*

The two most ardent supporters of this idea were a University of Nebraska professor named Samuel Aughey (who had worked with Hayden) and Bernard Fernow, who would become the first president of the Society of American Foresters. The public believed.

In 1873, Congress passed the Timber Culture Act, which made 160 acres available free to any settler promising to plant 40 acres into trees and maintain them for 10 years. Commissioner Wilson of the General Land Office

(GLO) had long instructed government surveyors to "plant midway between each pit and trench the seeds of trees adapted to the climate; the fact of planting and kinds of seeds to appear in the field notes of the survey" (in Hibbard, 412). With the passage of this new act, there was at last a mechanism in place that would have the settlers themselves do the work of turning treeless prairies and mountain valleys into forested bastions of America. And that "real" America was to be a humid landscape freed from heathen aridity.

The Timber Culture Act was largely aimed at the Great Plains, although "tree claims" were entered in the Rocky Mountain states. It was a law made to be evaded. The Act contained exemptions which made it possible for a person to receive a patent after only three years. However, many intentionally avoided gaining title since that would make them responsible for paying property taxes.

A common scam went like this: Buck Siemen enters a claim and quickly "relinquishes" it. A cohort who works for Buck then files on the same land and, a few years later (with no trees planted), gives it up. A second of Buck's ranch hands then files on it, and so on. Large land and cattle companies were assembled in this way. Since no money was invested, such groups received free grazing land or wheat fields for many years, and since the claims were never "proved up," they paid no taxes. It wasn't even necessary for claimants to reside in the state or territory where the land was located.

Speculation built up over the West like convective thunderheads. In 1878, the Timber Culture Act's requirements for issuing title to tree claims were amended by reducing the number of acres to be planted from 40 to 10 and the number of "living thrifty trees" per acre from 2,700 to 675. Most homesteaders viewed even those reduced regulations as bureaucratic nonsense. It was best to not bother with them. In 1887, the Commissioner of the GLO said, "I can truthfully say that I do not believe that one timber culture filing in a hundred is actually made in good faith for the purpose of cultivating trees" (in Hibbard, 419).

The Timber Culture Act quickly moved a lot of land into private hands, but it slowed real settlement by creating mounds of bureaucracy which fertilized land speculation. By the time of its repeal in 1891, 9,856,264 acres of the West had been "patented out." In the Rockies, the totals were small relative to the Plains: Colorado, 585,000 acres; Montana, 56,000 acres; Wyoming, 51,000 acres; Utah, 16,000 acres; New Mexico, 13,000 acres. The Homestead and Enlarged Homestead acts ended up patenting a hundred times that much in those states.

The notion of tree plantations as a source of wood and shade and as windbreaks would later be successfully advocated by the post–Dust Bowl Soil Conservation Service. This only came once the program gained a solid footing in rationality. There was no more talk of trees enticing rainfall from the sky.

In 1875, President Ulysses S. Grant, a man with a substantial thirst, was among the first influential people to acknowledge the Western landscape also required drink to be settled down. Grant said,

> *In Territory where cultivation of the soil can only be followed by irrigation, and where irrigation is not practicable . . . land must be held in larger quantities to justify the expense of conducting water upon it to make it fruitful. (in Hibbard, 425)*

General Grant knew that settlers needed more land than a 160-acre claim. In 1877, with growing public displeasure over the implementation of various homesteading bills, Congress passed the Desert Land Act. Under this bill, a settler and "his wife" could file on 640 acres in eleven western states and territories by paying 25 cents per acre. Title was issued once an additional $1.00 per acre had been paid, and the applicant "proved" irrigation water had been applied.

This act was an engraved invitation to speculators who RSVPed with great gusto. Vast areas of land were accumulated by cartels which shifted paper ownership from one member to another to keep one step ahead of the General Land Office. Such deceptive schemes were absurdly easy to maintain. Often the only irrigation came in the form of a bucket of water dumped sarcastically on the land in the presence of a witness. A Montana rancher once recalled a drunken party held by his grandfather on a Desert Land Act claim. His granddad and various laughing cronies sat around swilling whiskey and answering inevitable calls of Nature which were later solemnly attested to as "applied irrigation water."

Of the 32 million acres of original entries under the Desert Land Act, only 8 million acres were finally patented to settlers. The remainder slipped away, reverted to government ownership for later disposal, or became part of what are now Bureau of Land Management (BLM) holdings.

Major John Wesley Powell is best known as the one-armed stalwart of hair-raising explorations of the Colorado River watershed. However, his lasting mark was made not as an adventurer who confronted horrifying rapids, but as a scientist who squarely faced the fact of Western aridity. For a time, Powell held the title of United States Geographer and would be the first head of the U.S. Geological Survey. He possessed a stunning clarity of vision in the field, if not in the treacherous halls of the Capitol.

Powell saw endless land speculation as an avoidable waste of people's skill and heart. He was a proponent of developing the West's water resources for irrigation, but a voice of reason in the matter of scale. In the 1870s, the Major studied Mormon Utah and Hispano New Mexico, where irrigation on small tracts was already successful, in hopes of applying these principles throughout the arid West. He advocated land-use mapping and the distribution of private property based on watersheds, landforms, and access to irrigation water. While the Desert Land Act called for 640-acre allotments, Powell steadfastly argued that at least four full sections of land (2,560 acres) were needed per settler, with 80 acres of this being irrigable cropland. His 1878 *Report on the Lands of the Arid Region of the United States* contained reasoned recommendations on how the West might be rationally developed and brought into the national fold.

Powell was an advocate of uniformitarianism where aridity was concerned. He was the Lyell of Western climatology. For the well-traveled General, plowing and planting could not alter precipitation regimes. Leibig's Law of the Minimum states that the limiting factor in a system controls the system. Powell saw that water in the West was both limited and limiting. And while believing in water development, he advocated a far lesser density of population than the political times clamored for. As a result, he was widely attacked by politicians, land speculators, irrigation companies, and the public and was eventually dismissed as a naysayer in part because of an inability to sell his message back in Washington. Bernard DeVoto once wrote, "Of the Americans, it was the Westerners who first understood that there are other limits than the sky" (DeVoto, 364). Powell grasped the region's unrelenting central truth. And then, as now, in this contrary land, to speak of limits can bring on personal ruination.

The disposal of the West continued. The 1878 Timber Cutting and Timber and Stone acts in effect gave Western timber interests carte blanche authority to cut and run. Logging companies searched out high-grade timber, provided people with funds to buy a Timber and Stone Act claim for that parcel, then quickly bought the timber and land from the claimant. These sorts of tap dances around the law, together with widespread timber poaching on public lands, were early examples of the environmental devastation which is the signature of the Rocky Mountain timber industry to this day.

The establishment of Timber Reserves in 1891 and the formation of the U.S. Forest Service in 1906 briefly resulted in improved management, but excessive clearcutting is now once again destroying immense acreage of old-growth ecosystems. Today, the government subsidy is covert. Deficit sales (timber sales which lose Federal dollars) are the rule in the Rocky Mountain states. The impact on threatened and endangered species, such as the spotted owl in the Pacific Northwest, is receiving escalating national attention.

The Newlands Act of 1902 established ways to finance western water development schemes and provided the rationale for the Bureau of Reclamation. Thoreau once said in *Walden*, "If there is magic in the world it is found in water." If there is profanity in the West, it is found in the dam projects of what is sometimes called the "Bureau of Wreck the Nation." The Glen Canyon Dam/Lake Powell debacle on the Colorado is but one of scores of ill-conceived water "development" efforts which have turned many Western rivers into tamed ditches.

Yet, these massive water control schemes were precisely what most Americans wanted so that land development could proceed. During the succeeding years, a flood of bills continued to be passed by Congress which, despite diverse purposes, essentially served to give away the public domain in hopes of "building up" the Western United States.

In his book *To No Privileged Class*, Utah historian Stanford Layton analyzed the Enlarged Homestead Act of 1909 and the Stock Raising Homestead Act of 1916, as well as the social movements behind them. The Country-Life Movement arose in the Progressive Era and was largely inspired by President Theodore Roosevelt. His 1908 Commission on Country-Life (on which forester/conservationist Gifford Pinchot served) was chartered to broadly investigate the existing conditions of rural America.

The Commission produced three main recommendations: (1) resource surveys helpful to farmers should be done (which would lead eventually to the creation of the Soil Conservation Service), (2) agricultural extension services should be provided (and later were), and (3) programs of "social betterment" should be initiated. The Commission was composed of neo-Jeffersonian idealizers of the yeoman myth. According to Stanford Layton, they believed that "farming was a dignified and virtuous way of making a living and that ordinarily it was the environment best suited to the raising of responsible and well-adjusted children" (Layton, 10). Their report concluded that "not only in material wealth that they produce but in the supply of independent and strong citizenship, the agricultural people constitute the very foundation of our national efficiency" (in Layton, 10). Commission members earnestly believed it vital to protect and defend "a race of men in the open country that, in the future as in the past, will be the stay and strength of the nation in time of war and its guiding and controlling spirit in the time of peace" (in Layton, 10). President Teddy Roosevelt, no stranger to rural and remote places, chimed in: "No nation has ever achieved permanent greatness unless this greatness was based on the well-being of the great farmer class, the men who live on the soil" (in Layton, 10).

The Country-Life Movement expressed an ideology widely held in the nation, but never gained much momentum. However, it added to the social climate which led to the passage of the Enlarged Homestead Act in 1909. This was a dryland farming bill. It consisted of 320-acre grants of land containing no timber or mineral resources in areas remote from irrigation water. The impact was tremendous in the Rocky Mountain states.

This act's passage coincided with the rise of Hardy Webster Campbell as a dryland farming guru. The holy book of the Enlarged Homestead Act was his *Soil Culture Manual*, which offered the latest word on how to use a fallow period to grow grain in dry country without irrigation. Due to Campbell's innovation, the railroads, both land grant and non—land grant lines, began to heavily plat the prairies along their tracks as farmland and promoted isolated valleys as townsites. They often grossly exaggerated the agricultural potential of the land and became prime movers of sophisticated real estate speculation programs. Typical was the Great Northern Railroad, which attracted settlers to Montana's northern "High Line." A boom ensued as wet weather coincided with skyrocketing wheat prices in the early years of World War I. A bust soon followed as droughts and postwar economic decline set in. In the process, tens of thousands of settlers went broke and left. They had been railroaded.

The Back-to-the-Land craze came on the heels of the Country-Life Movement. L. H. Bailey wrote of the important differences between these two social phenomena:

> *They are not only distant, but in many ways antagonistic. The Country-Life Movement is the effort to make the real farming regions as progressive and as effective in a social and economic way as are cities and towns. The movement is thoroughly sound, because any effort to increase the efficiency of an existing civilization is sound. The Back-to-the-*

The Rocky Mountain West

*Land movement is the effort to place city and town people on farms. It is
in part an effort to relieve city congestion, in part an expression of the
desire of city people to escape, in part an effort of real estate people to sell
land. (Bailey, 379)*

The rush to return to the land gained momentum among the educated.
Greenhorns could sit in their easy chairs and fantasize over articles in *Colliers*
and *Sunset* which rhapsodically described the joy, leisure, profit, and improved health which awaited outside the city. As in the time of Jefferson,
farm work was again portrayed as a sacrament. Across the country, orchard-
ing was employed by real estate developers and water companies as a way to
lure naive urbanites into sure-lose gambles.

The Back-to-the-Land movement was felt in Washington. In 1916, nearly
40 years after Powell's *Report on the Lands of the Arid Region of the Western
United States*, Congress passed yet another bill to build up the West. The
Stock Raising Homestead Act provided 640 acres of grazing land to a settler
for $1.25 per acre. While one "section" in the West had already been shown
to be totally inadequate to support a family, this act succeeded in moving
huge amounts of land into the private sector. Congress was unconcerned
that many "settlers" were actually real estate companies. By 1934, some
60 million acres were dispersed under this little-known and relatively recent
homestead act.

The 1934 Taylor Grazing Act brought down the curtain on the main era
of governmental disposal of the public domain. Government land which had
not been distributed or incorporated into the holdings of Federal or state
agencies was then made available to the public via inexpensive grazing leases.

In 1946, a widespread Western land grab was attempted. The American
National Livestock Association and the National Woolgrowers Association
took their leads from Nevada Senator Patrick McCarran and pressed for sales
of 142 million acres of public domain at prices as low as nine cents per acre.
While some "surplus" land was sold, this early "sagebrush rebellion" largely
failed. They had reached too far. It was the same strategic mistake President
Ronald Reagan and Interior Secretary James Watt made during the attempted Assets Management land disposal program of the 1980s.

The Bureau of Land Management (BLM) was formed in 1946 by the
merger of the General Land Office, which had administered the various
homesteading acts, and the Grazing Service. The BLM continues to lease
public land for livestock grazing, often at a fraction of the open market price.

This subsidized provision of forage may be viewed as a continuation of the process of land and resource disposal.

In the 1950s, suburbs began to spread across agricultural land around the periphery of Western cities. Ironically, much of the financing for these homes came from the Farmers Home Administration (FmHA). FmHA lending regulations required that homes be built many miles from cities and towns. Instead of fostering the agrarian way of life, this Federal agency has been instrumental in the subdivision and suburbanization of farm and ranch land.

The 1960s had its Back-to-Nature movement, which had been foreshadowed 50 years before. The 1970s was a time when the song "Rocky Mountain High" was on everyone's lips and unprecedented profits were in developers' pockets. The Rocky Mountain states responded to these pressures by passing land subdivision laws and regulations. However, decades of viewing the land as a hot commodity or a sacred freehold would not be easily turned around.

Evasions and exemptions ruled the day during implementation of homestead bills. This shuffle continues today during the application of land-use planning regulations. In fact, planners estimate that 75 to 95 percent of all land splits in the Rockies occur with no planning office review. Therefore, the existing land subdivision laws in the region function as *de facto* homestead acts.

In Montana, landowners can sell off an unlimited number of 20-acre tracts without review. In Colorado and Wyoming, parcels of 35 acres and above are exempt. Most Rocky Mountain states' exemptions, some would say "loopholes," allow one tract per year to be split off without planning board oversight. In Utah, divisions of land for potentially dangerous commercial and industrial purposes can occur with no public hearing. Even in jurisdictions with well-trained planning staffs and environmentally aware citizens, legal and political factors still prevent planners from contending with land development in any meaningful way. For example, in Missoula County, Montana, over 90 percent of all subdivided tracts created between 1973 and 1984 received no local review (Bugbee and Associates, 21). As growth increases, the situation only gets worse. Of the 18,336 acres subdivided between 1987 and 1991 in Missoula County, less than 1,000 acres (5 percent) were formally reviewed and approved through the regulatory planning process (Henderson and O'Herren, 23). The percentage of planned subdivisions is even lower in many rural counties of the West. Zoning is rare in the Rockies outside urban centers. As a result, today's land developers are filling the niche of the homestead era land offices, railroads, and real estate

cartels. Their work goes on essentially unimpeded. City and county governments are now actually encouraging this development through "Special
Improvement District" (SID) bond sales which subsidize developers in com-
pleting roads, sewers, water systems, and other infrastructure within subdivisions. Assuring that "improvements" within developments are technically
sound is wise, but too often developers leave jurisdictions holding the bag
for delinquent SID payments when projects don't sell.

In *The Legacy of Conquest*, Patricia Limerick wrote,

> *Americans saw the acquisition of property as a cultural imperative,*
> *manifestly the right way to go about things. There was one appropriate*
> *way to treat land—divide it, distribute it, register it. (ss)*

This is still mostly the way things are. In the homestead era, beneath all the
fine rhetoric, Congress gave implied permission for the West to be thrown
open to land privatization by anyone with the energy to grab it. Aridity was
seen as a fallen state, and meteorological mysticism was devised to fulfill the
collective desire for Western development at Midwestern densities. The salvation of the West was fabled as possible if enough people plowed, planted,
and prayed. The minority of the public who were outraged at the way land
speculation drove development never measurably slowed this focal activity of
Western life. Instead, people began to accept it as the norm and engage in it
themselves.

Today the region is even more infested with promoters bent on splitting
the land and filling it with people. This unbroken cultural momentum is a
diversified, formidable force. City and county land-use planners and conservationists working in land trusts have their hands full. They are wrestling
with the musculature of Western cultural and historical geography. While
money is certainly a central motivation of developers, it's as if the quiet of
the land bothers them, and they turn to the buzz and clatter of platted civilization to blot out the silence. To others, the powerful beauty of the landscape is a catalyst which stimulates intense life experiences and the urge to
save geographic rarities. Both options still lie before us. For in the West, you
can stand on a rimrock and see a hundred years in either direction, or you
can choose not to see at all.

WAYS OF SEEING THE WEST

In *The American West as Living Space*, Wallace Stegner evaluated the results
of the region's widespread land subdivision.

Land and Life

> *The pioneer farmer as Crevecoeur conceived him has a place in west-*
> *ern history, and as Jefferson's yeoman he had a prominent place in the*
> *mistaken effort to oversettle the West, first by homesteading and then by*
> *reclamation. (72)*

What factors of environmental perception caused and continue to perpetuate this mistake? Stegner uses the term "space" in his book title. This word also shows up in a collection of essays on the West called "That Awesome Space." In *Virgin Land*, Henry Nash Smith called the West "the vacant continent." In *The Nine Nations of North America*, Joel Garreau renames the Rocky Mountain/Intermountain West "the Empty Quarter." The intended or unintended image of the region conveyed by these writers is one of unoccupied, blank space—the void. "Space" implies a vacuum destitute of geographical content, or worse, an admirable quality in a closet. It is something which needs to be filled up. "Place," on the other hand, is defined by detailed landscape information and terrestrial emotions. A "place" is a whole. We can enter and change it into another kind of place, but we can never *create* place from nothing.

When we refer to land as "space," we risk conveying a lack of knowledge or care. Therefore, in matters concerning land, "space" is the state which precedes individual and cultural knowledge. Once we know and care where we are, we become placed. We say, "I can't place her face," as if that person was known to us only in situ and lost identity away from that context. Events "take place" in certain locales. Ranches are described as "The Smith Place," not "The Smith Space."

Curiously, conservationists and planners refer to undeveloped land as "open space." This phrase also perfectly describes the open corridors of a shopping mall. The term "open place" is perhaps no better, but the distinction between space and place is important. Geographic illiteracy has manifested itself throughout Western history as a pernicious drive to exploit a "spacious" land.

In this context, Garreau's "Empty Quarter" image is absurd. How can the Rocky Mountain West possibly be imagined as empty? Empty of what? The failed demographic dreams of the homestead era continue to fill the slumbering consciousness of realtors and Chamber of Commerce members. To many of these people, the region's resource limitations are nightmares best left unanalyzed. It turns out there is historical evidence and climatological reason why the West will never live up to its legend. Water shortages and destructive water development projects, ongoing losses of wildlife habitat and biological diversity, and the often moribund economies of most Rocky Mountain states reveal that perhaps the region is already too *full*. It is also a challenge to the human soul to dismiss a landscape as "empty" where grizzly bears, wolves, and elk move beneath mountain crests and a sky full of soaring birds.

The dry, fragile Rocky Mountain province has a fullness few ecoregions approach. Yet it has been perceived of and treated as wet, vacant, and vast beyond measure—more like the sea than the land. It was stripped of buffalo much as the oceans were nearly swept clean of whales. Both the West and the sea have been dumping grounds and arenas for ceaseless resource extraction. Throughout most of history, both have been places where no one was watching, and if they were, they wouldn't say a word.

Bernard DeVoto angrily called the West a "plundered province," while Montana historian K. Ross Toole spoke of it as a "resource colony." Today, the west has become a chic place for disgruntled urbanites to move to and unwittingly destroy. This is the most dominant and troublesome cultural inheritance of Westerners—the notion that the land is a boundless Klondike or cornucopia which can absorb all development without succumbing. Environmentalist David Brower summed up the attitude this way: "There were so few people and so much world—that if you failed *here* you could always go *there* and try harder" (Brower, 321). In *American Myth; American Reality*, James Robertson agreed with Brower: "Movement—physical geographic movement—is the magic which keeps expectations high; it fuels a belief in unlimited opportunity and unlimited success" (242).

For many people who move West, the expression "back East" is much more than a compass bearing. To return "back East" is to fail out here where it counts. It is to regress to a past, less promising way of life; to backslide, back down, reel backward, and fall on your backside. Success as a person and as a society is to go forward with growth. This has often been easier said than done.

Boom and bust is the persistent driving rhythm of the West. Repeated booms and busts have caused a cyclical stressing of land and people. Like soil that shrinks and swells with desiccation and drenching, the landscape has been stirred by each economic cycle, leaving an increasingly impoverished but fresh seedbed for delusions and earnest dreams to grow again. Things have been this way for a very long time.

Geologic time is measured largely by death. Each Era, Period, and Epoch is delineated on the basis of fossil evidence of the extinction of some lifeforms and the rise of organic novelties to replace them. Geographer Yi-Fu Tuan has written that the past recedes constantly and escapes recapture. Another geographer, David Lowenthal, has called the past "a foreign country." In the Rockies, where raw bedrock is so visually prominent, the past seems less elusive. Sometimes we can even hear its warnings as biological boom and bust whispers from ancient layered rocks.

CONCLUSION

The Rocky Mountain West is indeed searching country. However, most of our dreams of grandeur have been dashed because we have failed to comprehend the magnificence of the natural world directly in front of us. It is ironic that the Earth itself has been a faithful recorder of our struggle. The very aridity and ecological fragility we have so long sought to ignore have indelibly etched each conceit on the Western terrain. The tangible evidence of failed homesteading, overgrazing, rapacious mining, ruinous timber cuts, myopic land and water development projects, and the region's long-standing tradition of unregulated land subdivision surrounds us—as does the record of our honest attempts at belonging. The Western landscape is the region's most splendid and meaningful cultural artifact. It reflects our ideals and beliefs about who we are and reminds us of all that we have been.

The controversy over the conservation of ecological, scenic, recreational, and historical resources in the Rocky Mountain West is basically a conflict between those who see the region as vacuous space and those who experience it as a bountiful web of places. In *The Interpretation of Ordinary Landscapes*, geographer Donald Meinig reminds us: "Thus we confront the central problem: any landscape is composed of not only what lies before our eyes but what lies within our heads" (34). In truth, the Rocky Mountain region contains many "Wests." Each of these landscapes, shaped by the flow of homesteading and settlement, manifests the spiritual geography of a distinctive culture searching to find its way.

The Northern Rockies have been called "The Native Home of Hope." Montana, Wyoming, and Idaho are stunningly beautiful and supremely difficult places to live. They share a general mercantile history. In all three states, after exploration and fur trading periods, Anglo settlers and the Army subjugated Indians to gain control of the land for mining, logging, farming, and ranching enterprises. Life in the Northern Rockies has proceeded free of a dominant religious order, with little modern ethnic struggle, with a relatively slow rate of population growth, and with only pockets of intense land development pressure. While these states have been treated as resource colonies by eastern industrialists, the short economic booms have most often been separated by extended periods of bust. Substantial acreages of biologically rich private land near Glacier and Yellowstone parks have been protected from future development by The Nature Conservancy, the Rocky Mountain Elk Foundation, government agencies, and the region's 10 land trusts. Recently, the challenge of conservation has grown more formidable. Californians have now found this once "hidden kingdom" and are arriving in unprecedented numbers. But there is still reason to be optimistic about the future of some of the most wild and supremely handsome landscapes in America.

New Mexico is a state fraught with ethnic divisions. While this contentiousness often breaks down along Anglo, Hispano, and Indian lines, the reality of life in the "Land of Enchantment" is lavishly cryptic. The state's turbulent history surrounding land tenure, dating back to the periods of Spanish and Mexican land grants, remains the central pivot on which modern life turns. In New Mexico, concerns over land title, social justice, and economic and cultural survival often overshadow the need to conserve landscapes. As a result, the state's 5 land trusts and several national conservation groups encounter significant resistance to land-saving in the state. However, their efforts are gaining greater traction as land trust members achieve a more sophisticated understanding of New Mexico's gloriously mystifying cultural geography.

Nowhere in the Rocky Mountain West does a greater contrast exist in cultural attitudes toward land and life than between secular Colorado and Mormon Utah. Curiously, these adjoining states are superficially alike in many ways. Both have rapid rates of population growth (doubling times of 25–30 years), dominantly Anglo citizenry, intense demands for land development, sprawling urban corridors centered around the region's largest cities, massive trans-basin water transfer systems, defense and high-tech industries, destination area ski resorts, national parks, and a wide array of magnificent natural features. However, Colorado and Utah are separate nations with profoundly different historical and cultural personalities. This divergence is starkly apparent in how the notion of landscape conservation is received in each state. Colorado has 27 of the Rockies' 43 land trusts and many very accomplished government open space acquisition programs. Utah has one fledgling land trust and no successful local government open space efforts.

Why does this astonishing difference exist? How are the private lands of Colorado and Utah being developed or conserved? And lastly, what do the vernacular landscapes of these neighboring realms reveal about people's internal worlds and their willingness to care about the everyday Earth around them? In the Rocky Mountain West—this strange and endangered searching country—these questions are reliable trails to the heart of the matter of land and life.

PART II
COLORADO: THE ICON

^^^^^

4

ELDORADO
IN THE ROCKIES

I never saw a country settled up
with such greenhorns as Colorado . . .
They thought they were going to have a second California.
—AUGUSTA TABOR, NINETEENTH-CENTURY MINER,
CAMP COOK, AND WIFE OF COLORADO SENATOR H.A.W. TABOR

The Great Plains in 1963. Back seat of the family Buick. Nodding off from the metronomic effect of hundreds of miles of wheat fields. Intense summer heat. No air conditioning. Sleeping.

Suddenly, you are nudged awake and an impossible mirage rises ahead. Is it clouds? No, it's snow! And mountains, unimaginably high and beautiful mountains! Smiles all around, then a foot to the floorboard.

Later that afternoon—the 11,316-foot summit of Fremont Pass above Leadville. Silence and gloriously cool, cool air. The sky so blue it seems black. Snowbanks still unmelted—in summer! Blue spruce and elk, marmots and columbine. A new world.

You dash upslope till the air runs out, finally stopping to pant and laugh and send hoots of sheer animal pleasure echoing off the crestline of the Continental Divide. Who would have dreamed a place like Colorado existed? Unbelievable. Perfect.

Such an experience is not unusual. A summer vacation crossing the high country by car or a ski trip to a resort is how most people first experience Colorado. Powerful memories and dazzling photographs are carried home to be shared with others. Through word-of-mouth and a crush of media-generated images, Colorado has become widely idealized. In fact, this Rocky Mountain state alone has been elevated to the level of a terrestrial icon, embodying all the glories of land and life in this astonishing region. In the

∧∧∧∧∧

collective national unconscious, Colorado is the Rockies archetype. However, as is often the case with such perceptions, precisely the opposite is true.

Colorado is spectacularly unique in nearly every respect relative to all other Rocky Mountain states. With some 3.3 million residents, Colorado has nearly twice the population of fecund Utah, which is the region's second most crowded state. While nearly identical in size, Colorado has seven times the population of adjacent Wyoming. Colorado has scores of 14,000-foot peaks. There are none in the rest of the Rocky Mountain West. Denver is a national city with a metro-area population of over 2 million. The Mile High City is the home of pro football's Denver Broncos, who have been in the Super Bowl three times since 1977. While they were crushed on each occasion, the team, and therefore the city, were beamed into 100 million American households during the broadcasts. No other Rocky Mountain state has an NFL franchise. Even the state's major league baseball team, the Colorado Rockies, sharpens the image. For most Americans, Denver is the only major city in the region which is not thought of as a hick town. Colorado also has Boulder, America's ultimate hip, mountain university enclave. Aspen and Vail are international draws for World Cup ski racers and playgrounds for the wealthy lifestyles of the rich and famous. The presence of Rocky Mountain National Park further reinforces the belief that Colorado *is* the Rockies.

An icon is a venerated image which becomes the object of uncritical devotion. Across the country, Colorado is often blithely believed to be a flawless environmental Eldorado where the dreary problems of ordinary life can be set aside. However, for those who actually move to the state, day-to-day life can be far different than what was expected.

Residents of other Rocky Mountain states find Colorado so atypical they often equate it with California. The analogy is not farfetched. The Centennial State is dominated by newcomers. Its economy is diverse and, despite lulls, is a regional powerhouse. The Front Range is supported by elaborate water delivery systems from faraway mountain basins. And water is in increasingly short supply. The Denver area is rife with urban decay, violent crime, crowded freeways, and horrific air pollution.

For 80 percent of Coloradans, life in their burgeoning cities along the Front Range is fast, expensive, and unstable. The state's extraordinary landscapes, from the Great Plains to the Rockies to the Colorado Plateau, are being converted to residential areas, recreational developments, gas stations, malls, burger stands, and ski lodges at an astonishing rate. The scale and pace

of land subdivision and development exceed that of any other portion of the
Rocky Mountain West.

Even the state's chief executive is atypical. Colorado Governor Roy
Romer once said, "Open space enables you to let your mind roll to think
about things you need to think about" (*Canyon Courier*, October 14, 1987).
In 1988, Romer attended the Land Trust Alliance's annual conference in
Estes Park. There had been a recent upsurge in land trust activity in the
state, and the Governor came to investigate. On the evening of June 23,
1988, at the old Stanley Hotel, Romer gave a speech unlike any ever given
by a Western governor other than, perhaps, Cecil Andrus of Idaho. Romer
asked:

> *What kind of wisdom have we? The beauty of the land is lost in our
> economy and when you're dealing with ecosystems, our actions are usu-
> ally irrevocable. We must help the free enterprise system measure the
> values we share. We need to lever it in. The economy of Colorado is com-
> pletely tied up in the state's beauty. I'm not talking about just wilder-
> ness, but about how we can use conservation easements to create open
> space links between communities and make a better world. For it is the
> whole nature of beauty which lies at the heart of a sense of community.
> And the people who live with beauty can relate to each other in beauti-
> ful ways.*

It is impossible to imagine conservative governors such as Bob Bennett of Utah or Marc Racicot of Montana delivering such a speech without large caliber guns pointed at their heads. Given Colorado's flair for
the unusual, from its governor to the intensity of development pressure,
it seems worth exploring how the land conservation efforts of the state's
land trusts and local governments are faring in several quite different sections of this fabled state.

The forces stimulating growth and the responses to it are as diverse in
Colorado as the state's extraordinary ecosystems. Twenty-seven land trusts
and numerous government open space protection programs are now confronting the loss of agricultural lands, wetlands, wildlife habitats, and historic resources (Map 3; Table 2). Along the Front Range, the development
tide sweeps in from many directions. Golden is absorbing the impact of suburbanization directly adjacent to Denver, yet substantial open space remains.

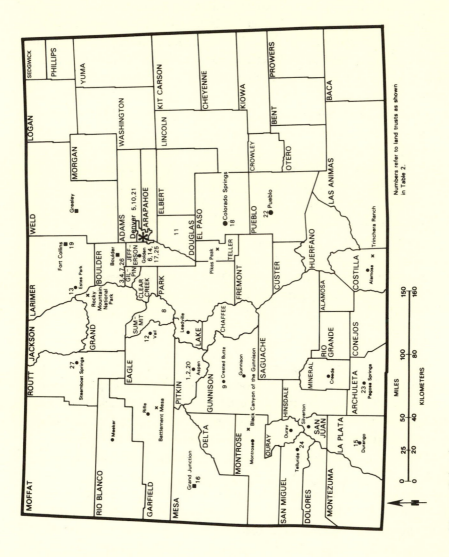

Map 3. Colorado

Colorado: The Icon

TABLE 2

Colorado Land Trusts

1. Aspen Center for Environmental Studies (Aspen, 1968: 475 acres)
2. Aspen Valley Land Trust (Aspen, 1966: 700 acres)
3. Boulder County Land Trust (Boulder, new group)
4. Boulder County Nature Association (Boulder, 1982: 184 acres)
5. Cherry Creek Trails & Open Space Foundation (Franktown, 1989: 0 acres)
6. Clear Creek Land Conservancy (Golden, 1986: 490 acres)
7. Colorado Open Lands (Boulder, 1981: 30,000 acres)
8. Continental Divide Land Trust (Breckenridge, new group)
9. Crested Butte Land Trust (Crested Butte, 1990: 40 acres)
10. Denver Urban Gardens (Denver, 1985: 2 acres)
11. Douglas County Land Conservancy (Castle Rock, 1987: 181 acres)
12. Eagle County Land Conservancy (Eagle, 1982: 60 acres)
13. Estes Valley Land Trust (Estes Park, 1987: 750 acres)
14. Jefferson County Nature Association (Golden, 1975, facilitator)
15. La Plata Open Space Conservancy (Durango, 1983: 1,070 acres)
16. Mesa County Land Conservancy (Palisade, 1980: 4,400 acres)
17. Mountain Area Land Trust (Evergreen, new group)
18. Palmer Foundation (Colorado Springs, 1977: 2,000 acres)
19. Poudre River Trust (Fort Collins, 1980: planning group)
20. Roaring Fork Land Conservancy (Aspen, 1989: 0 acres)
21. South Suburban Park Foundation (Littleton, 1979: 5 acres)
22. Southern Colorado Heritage Conservancy (Pueblo, 1988: 0 acres)
23. Southwest Land Alliance (Pagosa Springs, 1981: 2,175 acres)
24. Telluride Land Conservancy (Telluride, 1990: 0 acres)
25. Two Ponds Preservation Foundation (Arvada, 1990: 31 acres)
26. Wilderness Land Trust (Boulder, 1990: 310 acres)
27. Yampa Valley Land Conservancy (Steamboat Springs, new group)

Note: The location of the trust, its date of formation, and acres conserved as of 1/1/93 are shown in parentheses. Total acreage = 42,873.

Source: Land Trust Alliance 1991, phone survey December 1992.

Clear Creek and Gilpin counties are booming from tourist development spurred on by the legalization of gambling in several old mining towns. Fort Collins is a case study in planning and conservation in a community at the far edge of the Front Range corridor. And Boulder stands as a model of how environmentally concerned citizens and governments can create truly remarkable land conservation systems in the face of fantastic development pressure.

Colorado's Ski Country and the more remote west slope and southern regions also reveal much about land and life in the state. From mineral booms in the Piceance Basin to Western "chic" in Durango, to the fabled ski slopes of Aspen and Vail, Colorado's rural landscapes are also changing rapidly and under astonishing pressure to absorb even more development.

How did this seemingly remote and "unspoiled" state get into such a predicament?

This story begins not in present-day Colorado, but in the 1850s when it was first shouted that deposits of strange auriferous gravels gilded the surface of a distant, sunny land known as "Kansas." It is an appropriate beginning. For throughout history in this place born of boom, nothing has ever been done quietly.

5

THE FRONT RANGE: GEOGRAPHY AND DEVELOPMENT

In the early 1800s, the only outsiders the Arapaho, Cheyenne, and Ute Indians of the Front Range had seen were fur trappers, explorers, and a few wayward Spaniards. The quiet would not last. In 1858, the 103-member William Russell Green party discovered rich placer gold diggings along the South Platte River and its tributary, Cherry Creek. Two gold camps sprang up overnight in the plains at the foot of the mountains, and the gold rush to the "Pikes Peak Country" began.

St. Charles was established on the east bank of Cherry Creek, and Auraria (from the Latin root "aurum," which means gold) rose on the west bank. General William Larimer became an active promoter of the area and bought in at St. Charles. He renamed the camp Denver City in honor of Governor James Denver of the Kansas Territory, to which that portion of present-day Colorado belonged.

The John Easeler party soon set up sluicing operations at nearby Montana City and proceeded to split the land and sell lots. It is estimated that in 1859, 50,000 "fifty-niners" set forth for the "Kansas" gold fields. Certainly not all of them made it. However, enough did so that by late 1859 mining camps were scattered up and down the Front Range wherever creeks left the mountains and flowed across the plains.

In 1860, Auraria and Denver City consolidated to form Denver. For the remainder of the 1860s, the region was the domain of transient miners and speculators. In 1869, the Union Pacific laid track through Cheyenne on its

∧∧∧∧∧

way to Promontory, Utah, for completion of the nation's first transcontinental railroad. Denver was bypassed as much for the difficult mountain passes which lay to the west as for its small population of 3,500.

The 1870s saw the beginning of the metamorphosis of Denver into what historian Gunther Barth has called an "instant city." In 1870, when Illinois had a population of 2.5 million, Colorado had only 40,000 residents. Between 1870 and 1890, the population of Denver exploded from 4,759 to 106,713, making it the fastest-growing city in the United States for that period.

A number of factors contributed to this transformation. Mineral development spread into surrounding areas when the extraction of silver and lead became fabulously profitable. Agricultural colonies arose. Horace Greeley's 1870 Union Colony (which later became Greeley), the 1870 Chicago-Colorado Colony at Longmont, and the 1872 Fountain Colony at Colorado Springs were the most successful. The Kansas Pacific Railroad reached Denver from the Midwest in 1870, and a 106-mile branch line was extended south from Cheyenne shortly thereafter. The selling of Colorado up and down the East Coast as a tourist destination and clean-air health haven proceeded with evangelical fervor. By the late 1870s, William Jackson Palmer, founder of the Denver and Rio Grande Railroad, began advertising a "round the circle" tourist package consisting of a narrow-gauge journey through the "Alps of America." The Denver-based Colorado Mortgage and Investment Company and scores of land and water companies were actively marketing land to waves of newcomers. Coal trusts, flour trusts, and other cartels were established to assure high market prices for necessary urban goods.

By 1876, when Colorado gained statehood, Denver was the "Queen City of the Plains" and a major transport, banking, commercial, and tourist center. Inveterate English world-tramper Isabella Bird visited Denver in 1873 and described it as "the entrepot and distribution point for an immense district, with good shops, some factories, fair hotels, and the usual deformities and refinements of civilization" (Bird, 161). By 1890, Denver was home to over one-quarter of Colorado's 413,000 residents. In the coming decades, the demographic dominance of the capital city and the rest of the Front Range would continue to increase (Map 4).

By the turn of the century, Colorado's economy began turning away from mining toward agriculture. Between 1900 and 1920, over 200,000 people migrated to the Centennial State at a time when mining was in a slump. By 1920, Denver had become the core of a five-county metro area

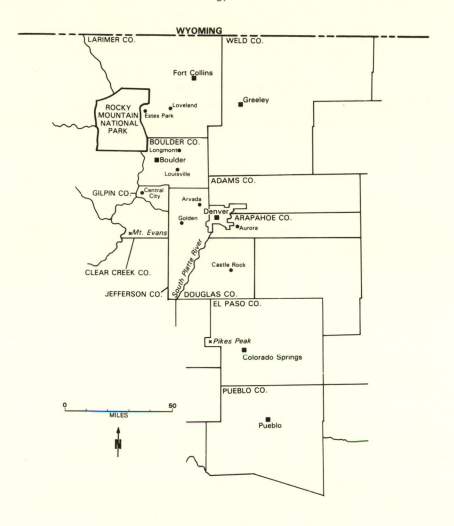

Map 4. The Front Range

The Front Range

with a total population of 331,000. Sprawl began to cover Denver, Adams, Arapahoe, and Jefferson counties. Aurora and Englewood, east and south of Denver, had become teeming suburbs with trolley lines linking them to the city. Car camping and other recreational pursuits attracted people to the mountains for weekend outings. Denver was a white city in those days. In fact, during the 1924 elections, the Ku Klux Klan allegedly helped elect the Denver mayor and police chief, as well as U.S. Senator Lawrence Phipps. After the 1920s, the Klan mostly faded from the Colorado scene.

In 1927, the 6.2-mile-long Moffat Tunnel was completed through the Continental Divide west of Denver. This provided direct rail service between Denver and Salt Lake City and on to the Pacific coast. The increasingly powerful Denver Water Board laid a pipeline with a 29,000-acre-foot capacity through the tunnel.

John Gunther once said that "water is the blood of Colorado." The Moffat Tunnel was the beginning of the transfusion of Colorado River water into the rapidly expanding Front Range urban corridor which by 1920 contained over half of the state's 1 million people. As early as 1921, the Denver Water Board had realized that the South Platte River would not be nearly enough to float its vision of eternal growth. It began filing claims on the Colorado River and analyzing sites for diversions under the mountains. In 1938, despite National Park Service objections, the Denver Water Board and the Bureau of Reclamation began the Colorado–Big Thompson Project. When it was finally completed in 1956, the $160 million delivery system consisted of thirteen dams, ten reservoirs (including the huge Granby Reservoir west of Boulder), canals, power plants, pumps, and a 13-mile-long tunnel through which Colorado River water was piped under Rocky Mountain National Park and gravity-fed to the Denver metro region.

Defense industries began to grow enormously influential in the Front Range piedmont zone during the 1940s. At the peak of World War II production, the Remington Arms plant employed 20,000 people, 40 percent of the Denver area's total factory work force. Postwar optimism fueled rapid urban transformation and suburban expansion. The population of the city fringe rose 73 percent during the 1940s, and by 1950 the five-county metro area had 612,000 people. The Bureau of Reclamation and the Denver Water Board completed the 23-mile long Roberts Tunnel to bring more water to the Front Range. Denver began to be vigorously promoted by Chambers of Commerce and business groups.

In 1948, Bernard DeVoto warned a University of Colorado commencement audience that the ongoing destructive exploitation of the state's resources would only lead to more complex and perhaps insurmountable problems. A few years later, historian Walter Prescott Webb wrote a piece in *Harper's Magazine* entitled "The American West: Perpetual Mirage." Webb used brutally plain logic to explain Western aridity and the limits it placed on urban growth. Coloradans scoffed at both DeVoto and Webb. In *The Coloradans*, Robert Athearn describes the *Denver Post's* sarcastic assessment of Webb:

> The Post *jeered at him, called him "Doc" as though he were some kind of quack and expressed great indignation at his suggestion that mountain-front urbanization was the result of flight from the desert. The title of the* Post's *article, "Us Desert Rats Is Doing Okay," was suggestive of the sensitivity felt by boomers and of a traditional inferiority complex that westerners still tried to cover with bombast.* (322)

As Dorothy Parker once said, "Ridicule is a shield and not a weapon."

The promotion and selling of Denver as a major American city attracted considerable media attention. The region received huge inputs of capital from Texas and East Coast investors, and national corporations opened branches in downtown Denver. Gates Rubber, a manufacturer of belts and hoses, became a significant Front Range employer. In 1956, Martin-Marietta chose the Front Range as the site where the Titan ICBM would be built. Seventeen thousand people were employed during the maximum bomb output period. The Colorado oil and gas industry grew as Americans began to drive more miles on the nation's expanding interstate highway system. The population of Colorado climbed 32 percent—from 1,325,000 to 1,754,000—during the 1950s. While growth in Denver itself was moderate, Adams County, northeast of the city, grew 200 percent during that decade.

In the 1960s, the Rocky Mountain lifestyle enticed government agencies and many coastal corporations to locate in the Front Range. IBM opened a plant outside Boulder which today employs 2,000 people. The same city was selected as the site for the labs of the National Bureau of Standards and the National Center for Atmospheric Research. Oil companies expanded operations in the region. At the Rocky Mountain Arsenal, next to Denver's Stapleton Airport, nerve gas production was stepped up. At the Rocky Flats Nuclear Weapons Center, between Denver and Boulder, Dow Chemical

The Front Range

Company ran the Federal government's expanded production line for plutonium triggers for nuclear warheads.

In Golden, west of Denver, the Coors Brewery became one of the region's major employers. However, despite Coors, Jefferson County had become a dominantly suburban landscape. From 1950 to 1970, its population increased 318 percent to 233,000. Over this same period, Arapahoe County grew 211 percent to 162,000 and Adams County soared 361 percent to 186,000.

From 1950 to 1970, the city of Denver became less and less residential and grew *only* 25 percent to 515,000 residents. During this two-decade span, the population of Colorado skyrocketed from 1,325,000 to 2,210,000 as the state's economy diversified into defense and high tech industries, oil, coal, uranium, and manufacturing. The normally reserved, analytical CBS News commentator, Eric Sevareid, became incensed over the rapid spread of houses onto ranch and mountain lands. In a 1962 article in the *Saturday Review*, Sevareid railed against realtors and concluded "to Hell with all of them."

The 1970s saw an even more extraordinary burgeoning of business which generated an ever-widening boom in the demand for residential real estate. The Johns-Manville Company relocated in Denver and bought the 10,000-acre Ken Caryl Ranch in the nearby mountains for the site of a corporate headquarters complex. The company constructed a $43 million business palace where clients were entertained in grand Rocky Mountain comfort. Visitors returned home to Chicago and New York raving about the new corporate/recreational lifestyle evolving along the Front Range. Honeywell, Monsanto, and Lionel quickly set up shop in the Denver metro area. Kodak built a plant in Fort Collins. In Colorado Springs, Hewlett-Packard, the Air Force Academy, NORAD, and Fort Carson combined to give that city a decidedly military and high-tech economy. Beech Aircraft and the National Oceanic and Atmospheric Administration built facilities in Boulder County. Greeley became the site of the world's largest expanse of stockyards. Agricultural processing plants opened in Longmont and Loveland. Many citizens of Front Range communities became concerned that their Eldorado was being "Californicated."

In response, the state legislature passed a basic land-use planning law in 1970. The measure called for each of Colorado's 63 counties to form land planning agencies. The bill allowed an exemption from planning review for parcels 35 acres and above. Environmentalists felt the law was intolerably weak and began to organize efforts to wrestle with "the growth beast."

Colorado: The Icon

In 1971, the Colorado Environmental Commission proposed legislation which called for a limit on the Denver metro-area population at 1.5 million and for the creation of an encircling greenbelt of parks and open space. The legislature never passed the bill due to questions about how it would be implemented.

Meanwhile, Denver boosters had begun formulating a bid to host the 1976 Winter Olympics. Colorado environmentalists at last had a high-visibility single issue around which to rally. The Colorado Open Space Council strongly opposed the proposal. They argued that the prospect of the Olympics would incite a land rush even more frenzied than what was already taking place.

Colorado politics were being changed by the environmental concerns of the very newcomers who were imperiling the state's ecosystems. In 1972, Pat Schroeder, a pro-environment Democrat, became the first woman to represent Colorado in Congress. Democratic gubernatorial candidate Richard Lamm rode point on the debate over natural resource issues. During his campaign, Lamm addressed a group of land developers and told them that if he was elected, they would find themselves "on the outside looking in." Unimpressed with Lamm's threat, land speculators had begun buying up large tracts near possible venue sites in the mountains. Olympic Realty, Olympic Land Sales, and Olympic Properties were a few of the instant companies poised for the kill.

Richard Lamm's rhetoric out on the stump was straightforward and emotional. He heralded the arrival of widespread land-use planning and told crowds "we're not going to let the exploiters rip us up and rip us off." Lamm won election based on this kind of blunt talk and an unflagging opposition to the Olympic bid. The 1972 statewide referendum on the Olympics defeated the proposed Colorado games 537,000 to 359,000.

The first half of the 1970s was a time of new faces in government and of bolstered optimism that the state's mounting environmental problems could be solved. Open space protection programs were created in a number of cities and counties. The so-called "Billboard War" broke out over roadside aesthetics. In downtown Denver, seedy Larimer Square was eyed as an historic district worth restoring. Novelist James Michener would write *Centennial* and call Colorado "*The* Place." Things were looking up.

However, the Energy Crisis of 1974 was a catalyst for an even greater upsurge in business investment in Denver and across Colorado. The state's plentiful supplies of oil, natural gas, coal, uranium, and oil shale led boosters to call it "America's Persian Gulf." Arco, Amax, and other energy corporations set up headquarters in Colorado—many of them in downtown Denver. Yet during the 1970s, Denver itself had begun to shrink.

By 1974, 40 percent of Denverites were black or Hispanic. School busing programs began. White flight led to a 1974 state constitutional amendment which expressly forbade Denver from annexing land or developed suburban neighborhoods around its margin. As the downtown was filled with new corporate skyscrapers, surrounding counties, such as Jefferson and Arapahoe, became even more suburbanized, and some Denver neighborhoods became crime-ridden ghettos. However, the money from the energy boom was growing. Even Governor Lamm ended up backing a plutonium enrichment plant in economically struggling Pueblo.

The growth juggernaut rolled on. In 1978, Denver's infamous "brown cloud" led to the city's air being ranked second only to that of Los Angeles in foulness. Residents lived in a spreading, sickening pall of auto exhaust fumes, manufacturing and power plant emissions, and smoke from residential wood stoves. The economy was expanding at a fantastic pace. More and more newcomers arrived in pursuit of high paying jobs in the oil industry or with one of the state's 350 high tech firms.

Aurora, east of Denver, was the fastest growing middle-sized city in the U.S. during the 1970s. Lakewood, west of Denver, was the third fastest. The entire Front Range urban corridor was expanding at staggering rates. During the 1970s, Fort Collins and Colorado Springs were, respectively, the fourth and ninth fastest growing cities in America. By 1980, over 2.4 million people lived along the base of the Front Range, with 1.6 million of them within the Denver metro area.

The Federal bureaucracy was also fattening. By 1980, over 33,000 Federal employees were working for scores of agencies inhabiting offices along the Front Range. Only Washington, D.C., had more. Denver was now not only the Queen City of the Plains but also the capital of the energy-rich Mountain West. The downtown had been transformed into a shining city of new steel and glass skyscrapers built by the oil companies. Executives sat above the world in an immaculate vertical urban enclave which came to be known as "the Oil Patch." It was largely the realm of Texans, old Colorado hands, and wildcatters who got lucky. Denver in 1980 was a town where the "oil bidness" was wide open, and it was possible to make and lose a fortune before lunch. The boomers were riding high.

Colorado: The Icon

Governor Lamm then initiated the Front Range Project as a forum for discussions about how communities might deal with the extraordinary growth in land development. The forum concluded that preserving open space was critically important throughout the entire corridor—from the Wyoming border to Pueblo. During the same time, a group of Front Range CEOs, known as the Colorado Forum, began to discuss transportation issues, water development, the future of Stapleton Airport, and other matters. They concluded that land conservation was beneficial to the state and to their long-term business interests. In 1981, as a result of the Front Range Project and the momentum of the Colorado Forum, a statewide land trust was formed and took the name Colorado Open Lands (COL).

The need for COL was clear. At the time, Colorado was losing 117,000 acres of open space and wildlife habitat each year. A 1981 COL pamphlet outlined the scope of the problem:

> *Colorado is growing at three times the national average. Eighty percent of the state's population is concentrated in the thirteen counties comprising the Front Range. The equivalent of one-half of all development that has taken more than a hundred years to evolve will be added in just twenty more years.*

The members of the Colorado Forum agreed with the private, free market approach to conservation COL advocated. Ironically, the only corporation to drop out of the Forum was Kodak, the processor of billions of prints and slides taken by recreating tourists.

During the mid-1980s, the oil boom produced a glut of crude which pulled Denver into an economic recession. The *Christian Science Monitor*, commenting on this slide, carried an article entitled, "Energy boom gone bust overshadows efforts to restart Denver's economy." However, at the same time, an economic analyst, Tucker Adams, of United Banks of Denver, wrote that the Denver economy was still growing. The growth in jobs during the 1970s was 5.4 percent each year. By 1985, it had dropped to a still healthy 3.7 percent annual rate. In overbuilt Denver, the slowdown brought a high vacancy rate for downtown office space.

However, throughout the slump the economy was diversifying. In the middle years of the Reagan era, Martin-Marietta received huge defense contracts which provided mega-profits in the armaments sector of the Front

Range economy. The Technical Products Division of Data General moved from Boston to Denver, citing the city's location, highly educated work force (one in four Coloradans is a college graduate), and the amenities of mountain life. Throughout the 1980s, tourism pumped an average of $1.5 billion per year into the economy of the Denver metro area.

In 1985, Denver's worsening air pollution provoked action by the Environmental Protection Agency (EPA). The city's veil of carbon monoxide had grown too obvious to ignore. Denver was given until December 31, 1987, to bring its air up to Federal standards. Otherwise, EPA threatened to withhold $30 million in Federal highway funds. Despite the city's eventual failure to meet air standards, those drastic measures were not invoked.

EPA did flex its muscles in the hotly debated Two Forks Water Development Project. In 1980, the Denver Water Board and 42 other Front Range cities, towns, and counties proposed to build a dam south of Denver at the confluence of the South Platte River and its North Fork. In 1987, the U.S. Army Corps of Engineers completed a $30 million draft Environmental Impact Statement (EIS) on the $500 million project. EPA rejected the EIS as inadequate and alleged that the Army Corps of Engineers purposefully omitted important information on the impact of the project on water quality and wildlife.

On August 30, 1989, EPA finally killed the Two Forks scheme. The Federal government cited excessive environmental costs both in Colorado and in downstream Nebraska, where the reduced flows of the South Platte would have seriously damaged critical wetland habitat for sandhill cranes and dozens of other waterfowl species. Melinda Kassen of the Environmental Defense Fund applauded the decision in print saying, "There are less costly and less destructive ways" to provide water to Front Range metropolitan users. For the first time, the sustainability of Colorado's continuously expanding instant cities had been officially questioned. In time, however, Front Range communities will undoubtedly resuscitate the Two Forks idea.

Boomers have recently gained ground in two other arenas. A massive new regional airport is being built near Denver to replace Stapleton. Four airlines will use it as their national hub. Also, the people of Colorado, who so strongly rejected the notion of hosting the 1976 Winter Olympics, came out in clear support of an attempt to bring the 1998 games to the state. Salt Lake City received the nod as the official U.S. entry, but Denver will certainly try again, now that public sentiment is behind the idea.

Yet state residents have counterbalancing impulses as well. Since 1983,

30 percent of the proceeds from the Colorado Lottery have been allocated to three functions. Money from a lottery conservation trust fund has been directed to more than 400 jurisdictions for park and recreation projects. Also, one-third of the budget of the Colorado Division of Parks and Outdoor Recreation comes from the lottery and is used to manage 31 state parks. Lastly, money goes into a capital construction fund for public facilities such as schools, fairgrounds, prisons, and state buildings.

Over $200 million has been generated for these purposes. The motto of the Lottery Commission is "for a more beautiful Colorado." Gambling proceeds have assisted land protection projects such as the North Star Ranch in Pitkin County, the Boulder Creek Trail in Boulder, Roxborough Park in Douglas County, Horsetooth Reservoir Park at Fort Collins, Golden Gate State Park in Golden, the Platte River Greenway in Littleton, and the Garden of the Gods Reserve near Colorado Springs.

CONCLUSION

The Front Range urban corridor is the Rocky Mountain version of an East Coast megalopolis. The Colorado piedmont, where 80 percent of Coloradans reside, is today a nearly continuous sprawl 180 miles long and 25 miles wide. This 4,000-square-mile swath is only a fraction of the state's 104,000-square-mile area, but it draws a wholly disproportionate amount of water from both east and west slope rivers.

Two-thirds of the state's water falls west of the Continental Divide where only 20 percent of the people live. The Front Range receives over 400,000 acre-feet each year from west slope drainages—enough to cover the entire state of New Jersey with an inch of water. The region is wholly dependent on a maze of dams, pipelines, pumps, tunnels, and reservoirs which provide water for homes, businesses, factories, and agricultural operations. The Front Range is in this respect very similar to the Los Angeles basin. In both Colorado and California, widespread urbanization has been possible only through exotic, unrelenting applications of hydraulic and political engineering.

In *Instant Cities*, Gunther Barth wrote that Denver and other instant cities:

> *came into existence Athena-like, full blown and self-reliant. . . . A past did not exist, and the present was continually rendered obsolete by the change which sustained development. (xxvi, xxviii)*

The Front Range

Barth also believed such cities were only "instant" until:

> *the sudden wealth and population explosions spent themselves, and the*
> *equalizing trends of urbanization and industrialization drew them*
> *into the ranks of ordinary cities. (xxviii)*

This implies a discontinuity between frontier era booms and modern life. For Denver and the rest of the Front Range, that is completely untrue.

In Colorado, as elsewhere in the West, changing technologies have created novel economic opportunities which financed the construction of new instant cities atop the infrastructure of the older, equally instant urban landscapes. Development has spread up and down the Front Range and has interconnected cities and towns into an almost continuous smear of urbanization.

Tremendous losses of wildlife habitat, open space, agricultural land, and plant communities have resulted. However, in many jurisdictions, local governments and land trusts are attempting to conserve important portions of what is left. Given the tradition of explosive population growth and land consumption along the Front Range, how much conservation is actually taking place?

6

GOLDEN TO
FORT COLLINS: THE
TRIALS OF SUBURBAN
LAND CONSERVATION

GOLDEN

The gleaming silver Coors Brewery is the best-known landmark in Golden, the seat of suburban Jefferson County, lying west of Denver. Golden sits at the mouth of Clear Creek Canyon flanked by undeveloped basalt-capped mesas which rise above a sea of new residential housing districts. The cities of Arvada, Wheat Ridge, and Lakewood are aligned north to south along Jefferson County's border with Denver. The Colorado School of Mines is tucked west of the mesas near the massive brewery.

The county's 318 percent growth in population between 1950 and 1970 was the impetus for local conservationists to act. In 1971, a government organization called PLAN JEFFCO was formed to find ways to conserve the area's vanishing natural landscapes. By 1972, local voters had passed a one-half percent retail sales tax to raise money to establish the Jefferson County Open Space Program. All funds were to be used for open space planning, land acquisition, and property management. By 1992, the county had purchased and protected 22,000 acres in the mountains and adjacent to cities and towns.

The county's largest and most biotically diverse mountain reserve is the 3,040-acre White Ranch Park northwest of Golden. This easily accessible property supports elk, deer, black bear, mountain lion, and wild turkey. Miles of hiking trails make the site a superb day-use area within ten minutes of town. The nearby Elk Meadow Park, in the mountains south of I-70, protects 1,140 acres of elk range and other habitats.

∧∧∧∧∧

Golden is encircled by parks. Jefferson County owns a 110-acre parcel near Lookout Mountain with an historic mansion, conference facility, and nature center. South of the city, the Apex Park (530 acres) and Matthews/ Winters Park (1,095 acres) contain high quality foothills ecosystems. The Rooney, Foothills, and Civic Center parks comprise some 500 acres of developed recreational and open space land adjacent to the city. The Standley Lake Park, north of Arvada, preserves 465 acres of remnant prairie grasslands.

Perhaps the most intriguing county acquisition is the 18-acre Westchester Tree Grant Park located east of Golden in a suburbanized district. Originally granted to a settler named C. C. Welch in 1878 under the Timber Culture Act, the parcel contains many hardwood trees sprouted from the stumps or grown from the seeds of Welch's original plantings.

Jefferson County recently completed an updated open space and trails master plan to guide its efforts during the next 20 years. Two local land trusts will likely be there to help. One of these trusts has a long history.

In the mid-1960s, Carla and Patrick Coleman gazed east from their Lookout Mountain property to a spreading morass of suburbs, shopping centers, and highways and decided there had to be a better way. Despite the escalating price their land would bring on the real estate market, the Colemans held their ground. They soon formed Northwoodside, Inc., which Patrick Coleman has described as a "group organized specifically to protect wildlife habitat." In 1987, with assistance from the Trust for Public Land, this organization was restructured into the Clear Creek Land Conservancy (CCLC). The mission of the trust is to protect the spectacular Clear Creek Canyon from its mouth at Golden upstream 12 miles to its forks.

The Canyon is reminiscent of Black Canyon at Gunnison in western Colorado. Horace Greeley visited the drainage in 1859 and described it as:

> *a cold, swift, rocky bottomed stream, which emerges through a deep, narrow canyon. Brooks of the purest water murmur and sing in every ravine; springs abound; the air is singularly pure and bracing; the elk, deer, and mountain sheep are plentiful except where disturbed by the inrush of emigration; the solitude was sylvan and perfect until a few weeks ago. All now is being rapidly changed and not entirely for the better.* (New York Tribune, *June 9, 1859*)

With the publication of this account and his subsequent agricultural colonization efforts, Greeley became a major contributor to population growth and the very environmental disturbance he regretted.

Colorado: The Icon

During the nineteenth century, Denver and Colorado Springs led the way in conservation by forming parks at the ends of trolley lines and buckboard roads. Today, the Clear Creek Canyon is remarkable for the presence of black bear, elk, and eagles within sight of the Denver skyline. Clear Creek Land Conservancy members often describe the canyon as "a wilderness at the end of mass transit."

Thus far, the trust has conserved 490 acres in the canyon area, including a conservation easement on the Coleman's 240 acres lying along the canyon's south wall. This easement includes portions of the Beaver Brook Trail, Colorado's oldest documented hiking path. The trust also owns 10 acres fronting the trail and holds a lease with an option to buy 240 acres of Bureau of Land Management property.

More projects are underway, including a master plan for the canyon and several ongoing negotiations. In 1992, a major problem was averted. The owner of 9,000 acres near Clear Creek Canyon had planned to turn part of it into the largest open pit gravel mine in Colorado. The Jefferson County Commissioners voted not to grant the project the necessary permits.

The Jefferson County Nature Association, the county's second local land trust, operates out of a nature center atop Lookout Mountain. The group has been active in negotiations over wetland habitats such as Ralston Creek near Arvada, which contains some of the best riparian ecosystems left in the county.

The Nature Association also has worked with the Army Corps of Engineers in reviewing various applications for wetland alteration permits. The organization is conducting a countywide inventory of Great Plains riparian habitats and evaluating their condition and ownership status. Contacts have also been made with local landowners and the few ranchers who remain in the area. The group expects to gain in effectiveness once it receives training and technical assistance from the Trust for Public Land (TPL).

In 1989, a joint project of TPL's Santa Fe office, Jefferson County, and the Jefferson County Nature Association resulted in the purchase of a 79-acre parcel of pristine mesa-top prairie on Green Mountain near the Coors brewery. The property is now managed by the Jefferson County Open Space Program.

The Two Ponds Preservation Foundation operates out of Arvada. This three-year-old group helped secure a $500,000 Land and Water Conservation Fund appropriation which the U.S. Fish and Wildlife Service (USFWS) used to purchase a 23-acre wetland tract. This parcel was slated for filling and residential development. The City of Arvada recently donated a contiguous

8-acre tract, and the 31-acre preserve is now an urban national wildlife refuge managed by the USFWS. The trust is now working to expand the refuge.

Jefferson County, lying between the Denver skyline and the high peaks of the Front Range, reveals that development need not erase all of what a place was. Sometimes stands can be taken and held.

CLEAR CREEK AND GILPIN COUNTIES

Gold and silver ore deposits enticed people high into the Clear Creek watershed in 1859. Those who followed the creek's north fork created the towns of Blackhawk and Central City in what is now Gilpin County. Today, Central City has, according to historian Robert Athearn in *The Coloradans*:

> *turned into the Coney Island of the Rockies and the one-time Queen of the Little Kingdom has become a tired old bawd, painted up beyond recognition, selling her wares for any price to anyone. (323)*

In 1991, casino gambling came to the high country. Every building in the once weather-beaten downtowns of Central City and Blackhawk is now being repaired, renovated, or replaced to house slot machines, blackjack tables, and roulette wheels. Real estate prices have jumped tenfold and more. Crowds have already descended on the half-finished area to drink and gamble away spectacular sunny days. The sound of hammering, drilling, and cement trucks mixes with the dings and clangs of slot machines devouring quarters. Parking is nearly impossible. Shuttle buses are used to ferry in fresh reinforcements for gamblers too broke, beered up, or burned out to continue. Police cruisers haul off those who get unprofitably out of control. Many local people, who once enjoyed the solitude and safety of life in these small towns, have reluctantly sold out to the development corporations and moved on. Central City has become Atlantic City with alpenglow.

The original miners who moved up the South Fork of Clear Creek established Silver Plume, Idaho Springs, and Georgetown as the nucleus for Clear Creek County. Georgetown, like Central City, depends on tourism. Its mining-era Hotel de Paris has been converted to a museum. Signs on I-70 attempt to entice motorists to exit and tour the collection of rustic leftovers. The Geneva Basin, Loveland Basin, Berthoud Pass, and other ski resorts also attract Front Rangers to the two-county region.

Overwhelmed by these trends, Gilpin and Clear Creek counties have

Colorado: The Icon

now become a regional center for recreational and residential homesite development. However, the area is also the setting for one of the West's most innovative and heartening land conservation projects, whose roots lead back to the Civil War era.

In 1861, Abraham Lincoln appointed William Gilpin as Colorado's first territorial governor. Within his first year in office, he took some questionable drafts from the U.S. Treasury and was purged by the president. Illinois native John Evans was selected as Gilpin's replacement.

Governor Evans assembled a large ranch in present-day Clear Creek County at the base of the 14,264-foot peak which now bears his name. After his death, Evans's heirs continued to own the property under a family trust arrangement. By 1980, suburbanization had begun to envelop the ranch, and money was being lost on the agricultural operation.

Since the terms of the family trust required that revenue be generated for the heirs, the courts ordered the trust broken and the ranch sold. This placed one of the Front Range's main elk calving grounds in jeopardy. The property lies only forty-five minutes from Denver, and numerous developers poised for action. Under existing county zoning, 1,700 housing units could be built on the 3,200-acre ranch. Subdivision and development seemed inevitable.

In 1981, Colorado Open Lands (COL) stepped into the picture. The trust was staffed by two talented but dissimilar individuals: Marty Zeller, a thin, blonde man with sharp features and an M.A. in Planning from Harvard, and 6'6" Don Walker, who came to Colorado in the 1950s on a basketball scholarship to the University of Colorado in Boulder.

When COL was formed in 1981, the oil and energy business was in a deep economic recession. Don Walker recognized that the group must be self-supporting—that charity alone wasn't going to get the job done. The trust assembled a diverse board of directors consisting of public leaders and private entrepreneurs. Included were the chairman of the State Bank Board, the vice president of a huge Front Range development known as Mission Viejo, and the president of the sprawling Inverness development.

What COL really required was a successful project on which to build credibility. The impending sale of the Evans Ranch was, to Walker, the perfect bloody baby seal for the cover of their pamphlet. Walker and Zeller devised an audacious plan to save the property.

The courts had ordered the ranch liquidated, but the heirs wanted to conserve as much land as possible. This provided a wedge of daylight for

COL. Walker and Zeller approached the officer of the family trust to discover if any flexibility existed. According to Zeller, they found out that what he really wanted was money—plain and simple.

The COL board and staff decided a partial development design was the likely way to finance conservation of the ranch. The property's price tag was $4,665,000. COL didn't have one-tenth that amount in its endowment fund. What was clearly needed was an enactment of the old loaves and fishes parable. In a remarkable display of chutzpa, COL bought a one-year option on the 3,200-acre ranch for a mere $5,000. In twelve months, the fledgling land trust would need to make a payment of $445,000 or lose its $5,000. Pressured by the family, the bank holding the note on the ranch agreed to this seemingly futile plan.

To design a partial land development project, COL turned to planner Ian McHarg from Philadelphia. McHarg is best known as the author of *Design with Nature*, which influenced a generation of land-use planners and conservationists. After a few months, McHarg and the COL staff completed a plan they called "Stewards of the Valley."

Under the "Stewards" plan, the property would be divided into five individual ranches of 640 acres. Each ranch would have a 35-acre building area, and the remainder would be set aside as permanent open space protected by a restrictive covenant held by COL. Every owner would have the right to hike and hunt throughout the entire open space acreage within the original Mount Evans property. The restrictive covenants on the ranches could be converted to conservation easements whenever each landowner could use the tax advantages most beneficially.

The plan was set in motion, with a price tag of over $1.5 million for each of the five ranches. By then, the first payment on COL's option was coming due. The trust's operating funds were drying up. Walker and Zeller approached the Gates Foundation in Denver for a $5,000 contribution to cover day-to-day overhead. The Gates people made a surprising counteroffer. According to Walker, the foundation agreed to loan COL the money for any payments they couldn't make on the Evans Ranch. If they couldn't make any, then Gates owned the land. With no collateral, Walker and Zeller had eased away with a $4.7 million loan guarantee.

The Gates Foundation never got a chance to write any checks. In 1982, three of the five 640-acre ranches sold, raising $4 million for COL after the first payment was made. By 1987, the last two ranches were sold, and the Evans property was permanently protected. In the end, $3 million went into COL's endowment fund. The big, bold move had paid off.

Colorado: The Icon

However, land trusts have much more work awaiting them. Land values and the demand for recreational and commercial property are escalating rapidly as a boom arising from legalized gambling rumbles through Central City and Blackhawk and rolls across Clear Creek and Gilpin counties.

FORT COLLINS

Isabella Bird's 1873 Colorado sojourn was focused on the Fort Collins area, which later became Larimer County. The *New York Tribune* called this adventurous Englishwoman "a fair Amazon who rides like a Centaur over the roughest passes of the Rocky Mountains" (Bird, review on inside cover). Upon arriving in the military post known as Fort Collins, the saddle-sore Amazon began complaining at once, "This is a most unpleasing place . . . altogether revolting; entirely utilitarian, given up to talk of dollars as well as making them, with coarse speech, coarse food, coarse everything" (Bird, 39).

She was, however, enamored of the surrounding landscape:

> *Plains, plains everywhere, rolling in long undulations like the waves of a sea which had fallen asleep. It is covered thinly with buffalo grass, the withered stalks of Spanish bayonet, and a small beehive-shaped cactus. One could gallop all over. (Bird, 32)*

Fort Collins and Larimer County grew quickly based on agriculture, mining, and logging. By 1960, the county had 53,000 residents. Ten years later, it had grown to 90,000, most living in the Fort Collins/Loveland area.

Land-use planning was initiated in Larimer County in 1973 when the first comprehensive plan was begun. The Poudre River Greenbelt Association, the Task Force of Design Tomorrow Today, the Audubon Society, Trout Unlimited, and the Larimer County Horsemen's Association worked together from the beginning of the planning process to get local government to begin conserving natural ecosystems and open space.

They were successful. A huge seven-year capital improvement program was funded by local taxes, of which $2,350,000 was earmarked for an open space and trail program. Land was acquired in the foothills and along the Cache La Poudre River and its tributary, Spring Creek.

In 1978, the county comprehensive plan was completed, as was a plan for the town of Estes Park at the east portal of the 265,000-acre Rocky Mountain National Park. However, by the time the first county open space plan was written in 1980, Larimer County was in the throes of a monumental real estate boom.

The Fort Collins Standard Metropolitan Statistical Area was the fourth fastest growing such unit in the country during the 1970s, with an annual population increase of 5.4 percent. With the addition of nearly 100,000 new residents between 1960 and 1980, Larimer County had a pressing need for more open space and trails. The open space planners called for exactions from land subdividers for dedicated park sites and natural areas. In 1982, Fort Collins adopted the Land Development Guidance System to better evaluate a flood of development proposals.

This regulatory approach was based on a point system used to evaluate proposed projects. Review criteria were established, such as neighborhood compatibility, resource protection, environmental standards, and site design. Each criterion had a list of detailed questions about how well the proposed development met planning goals. A numerical rating was assigned for each concern. A "0" was given for a poor design, a "1" was given for an adequate plan, and a "2" was given for an exemplary job of meeting planning goals. The emphasis in implementing the guidance system was on a rational, predictable project ranking process and a willingness to negotiate. However, while the novel system resulted in responsible, environmentally based planning, it was not geared specifically to conserve land.

In 1982, the City of Fort Collins and the Loveland Chamber of Commerce asked Colorado Open Lands for advice about how they might secure more open space. COL proposed establishment of a self-funding transfer of development rights program. Development rights were to be sold by landowners in designated conservation zones to those owning property considered optimum for housing and commercial construction. The government balked at this confusing and politically dangerous scheme, and momentum quickly waned. Local voters had already voted in a sales tax to finance acquisitions, and the feeling was that this was enough. In 1985, an increase in the "open space" tax was proposed and rejected.

The Poudre River Trust, named for the Cache La Poudre River, which flows through Fort Collins, then prepared a policy plan for the urban riparian corridor. The trust is actually a planning advocacy group with no real estate functions. In 1988, Fort Collins followed up by preparing a parks and recreation master plan. Open space target areas included the hogbacks west of the city and the floodplains of the Poudre River and Fossil Creek. Today, the city has secured 17 miles of trails and 13 open space sites totaling 978 acres.

Colorado: The Icon

The 1990 population of Larimer County was approximately 200,000, four times the 1960 figure. With such a tumultuous onslaught of development, it is remarkable that any land has been protected at all. However, the sprawl continues to expand. Near Fort Collins on Route 287, developments such as Applewood, Clarendon Hills, Willow Park, the Woodlands, Wagon Wheel, and Horsetooth Commons contain some of the latest in downscale, pastel, patio living. One complex is using an old barn as a billboard which reads, "Five Oaks—Finally More Than A Townhouse."

Larimer County is also the locale of the worst "natural disaster" in Colorado history. On a summer evening in 1976, 60,000-foot thunderheads built up over Rocky Mountain National Park west of Loveland. The storm quickly poured a foot of rain into the Big Thompson River drainage around Estes Park. A mammoth flash flood charged down Big Thompson Canyon, tumbling 10-foot boulders, ripping out Route 34, and smashing houses into kindling. The torrent claimed 145 lives. Some torn and battered bodies were swept 15 miles to Loveland and deposited along the floodplain.

Today, Route 34 has been rebuilt, and people have constructed new houses in the canyon. The road provides the major access to Estes Park and Rocky Mountain National Park. The only concession to the disaster seems to be signs placed along the gorge urging people to "Climb to Safety in the Event of Flash Flood!"

Estes Park was the ultimate travel destination of Isabella Bird. She was enraptured by its wild scenery:

> There were chasms of immense depth, dark with the indigo gloom of pines, and mountains with snow gleaming on their splintered crests, loveliness to bewilder and grandeur to awe, there were still streams and shady pools, on the mountains dense with pine were patches of aspen which gleamed like gold. (Bird, 90)

Bird welcomed the quiet of the remote valley. She said, "Solitude is infinitely preferable to uncongeniality and is bliss compared with repulsiveness" (Bird, 157). It's a good thing Ms. Bird is presently on an adventure in the afterlife: Estes Park is now a ghastly tourist town complete with fast food franchises and chic clothing outlets.

The town is the base of operations for motorists touring Rocky Mountain National Park along Trail Ridge Road. The massive, white Stanley Hotel stands high above the din. It was built in 1903 by Massachusetts native

F. O. Stanley, who was the co-inventor of the Stanley Steamer. Portions of the Jack Nicholson horror movie "The Shining" were filmed in the hotel.

In 1987, local people set up the Estes Valley Land Trust to counteract the surge in recreational subdivision. A land trust member acknowledged, "We've got our hands full—it didn't get this way in a couple of months, and it will take years to make a dent in the problem here." Against considerable odds, the trust has conserved 750 acres of ecologically sensitive open space.

County planners project the population of Larimer County in 2010 will range from 300,000 to 500,000. The Nature Conservancy recently bought a 1,128-acre ranch and secured a 480-acre conservation easement in Phantom Canyon on the North Fork of the Cache La Poudre River northwest of Fort Collins. In Larimer and all of the growth-plagued counties of the Front Range, each such project is nothing less than critical.

7

BOULDER:
A MODEL OPEN
SPACE COMMUNITY

Boulder is a city lying northwest of Denver that tilts toward the far side. It has been the home of 1972 Olympic marathon champion Frank Shorter and one-legged world-class cyclist Andy Pruitt. Mork from Ork landed there. It is also the setting for the most innovative and successful collection of land conservation programs in the Rockies.

Settled by a group of Nebraskan gold miners in 1858, Boulder is now a rampagingly popular city of 85,000, of which 22,000 attend the University of Colorado. Isabella Bird passed through the "Boulder tract" in 1879 and reported, "Boulder is a hideous collection of frame houses on a burning plain" (Bird, 230). Things have changed.

PEARL STREET MALL

The Pearl Street Pedestrian Mall in downtown Boulder was completed in 1977 and has been this environmentally conscious community's focal point ever since. A small stream flows past tree-shaded flower beds, lawns, and metal statues. One of the statues, entitled "Hearts on a Wing," is of a young, long-haired woman in a pastoral summer dress sitting barefoot on a park bench. People often put real flowers in her sculpted hair.

Antique street lights adorn the rows of clothing boutiques, bookstores, galleries, sporting goods shops, bars, and restaurants. During lunchtime in Boulder, cuisine from any of the Earth's continents, except Antarctica, can

∧∧∧∧∧

be sampled. The businesses range from the Brick Shirt House (one T-shirt says "Beer and Loafing in Boulder") to the New York Deli to Santa Fe Ambience, which advertises itself as "an eclectic southwest gallery of sorts." Just off the mall is the Boulder Theatre, a 1936 Art Deco showplace restored in 1988. Nearby is a health food market called Alfalfa's.

Pearl Street is superb for people-watching. Students with backpacks dash around between classes at the university. A ranch couple sits eating ice cream cones and listening to a Peruvian Indian band. World-class triathletes talk with street rappers. A magician entertains the shoppers by escaping from a strait jacket while singing the old Patsy Cline song "Crazy." Fashions range from zoot suits to hiking boots. Cobalt blue and jade green mountain bikes are locked in racks atop BMWs parked next to neo-hippie VW vans awash in Grateful Dead stickers. There is an airy easiness to it all.

The Pearl Street Pedestrian Mall is an exceptional piece of community development planning. Prior to 1977, businesses were fleeing to shopping centers on the fringe of town. Today, the Pearl Street Mall is a classic representative of the "rustic-chic" style of commercial development. It is the kind of hiking trail that might have been designed by Marshall Field, not Bob Marshall. Yet it gives Boulder a heartbeat. This sort of lifestyle retailing is now as Western as a hay baler, an oil well, or an atomic bomb. It is the West, not of myth, but of day-to-day function. For in the late twentieth century, shopping has replaced rodeo as the region's sport of choice.

HISTORY OF CONSERVATION

The young, environmentally concerned, recreating populace of Boulder County has long supported the conservation of open space lands. The City of Boulder and Boulder County have assembled remarkable open space systems which have attracted well-deserved national attention. The histories of these programs are interwoven. The first project started in 1898 when some Texas school teachers bought 80 acres as a site for their summer chautauquas. This site is now Chautauqua Park, lying below the steeply sloping red rock formations called the Flatirons on the city's west side.

In later years, the city acquired other "mountain park" lands near the Flatirons, which form the community's spectacular backdrop. By the late 1950s, local people grew concerned about the spread of development around the periphery of Boulder. In 1960, the report of Trafton Bean and Associates strongly recommended that Boulder County begin protecting open space.

As a result, the county formed a Regional Park Advisory Committee, hired a parks coordinator, and assembled a parks and open space committee.

The arrival of IBM in Boulder in the early 1960s created several thousand new jobs and intense development pressure on agricultural lands. In 1967, city voters passed an additional 1 percent local sales tax, of which 40 percent would be spent on acquiring, protecting, and managing open space lands both within and outside Boulder.

Both the city and the county began to acquire properties. In 1971, Boulderites wished to do more and passed an amendment to their city charter which allowed the government to issue open space bonds. By 1973, the city council acknowledged the success of the program with the establishment of an Open Space Board of Trustees.

Boulder County, not to be outdone, created its Parks and Open Space Department in 1975. In 1978, the city and county collaborated on the Boulder Valley Comprehensive Plan, which has since gone through several revisions. The 1978 plan focused on how open space protection could be used as a way to sculpt the pattern of urban growth. Along with these policies came the region's first open space maps.

The 1978 comprehensive plan was a vast improvement over the "spokes in the wheel" concept which guided development in the 1960s. The "spokes" plan encouraged sprawl along arterials and freeways. According to the 1986 plan update, urban sprawl "encouraged the proliferation of numerous quasi-municipal corporations which significantly add to the costs of providing governmental services and increase the inequities between various groups of citizens in the tax burdens they must bear."

Between 1959 and 1974, Boulder County led the state in the amount of agricultural land which was converted to residential and commercial uses. Many landowners sold out to speculators who then held the property for lucrative resale to developers. The 1986 Boulder County Comprehensive Plan stresses eliminating this hopscotch pattern and states, "Open space should be promoted as an urban-shaping method." "To this end," said a planning official, "the joint plan has focused on Community Service areas as a way to encourage future development within existing urban centers such as Boulder, Longmont, and Broomfield."

CITY OF BOULDER OPEN SPACE PROGRAM

The City of Boulder's Open Space Program is both an effective conservation real estate brokerage firm and a land management agency. To date, the city

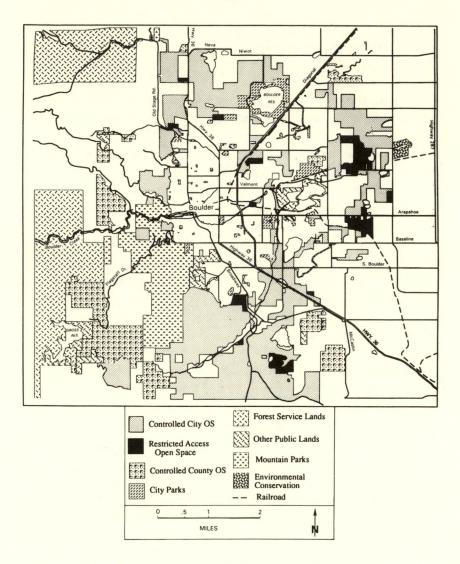

Map 5. Boulder Open Space (Source: Boulder County, 1992)

Colorado: The Icon

program has protected 34,000 acres at a cost of around $79 million. The Open Space Department manages 27,000 acres, and 7,000 acres are managed by the Mountain Parks office (Map 5).

Five city "Open Space Rangers" patrol a vast system of trails that wind through foothills, mountain meadows, and conifer forests. Large expanses of greenbelt open space frame Boulder's south side near the hamlets of El-dorado Springs and Marshall and parallel Route 36 as it approaches the city from Denver. Some 1,100 deer live within the greenbelt habitats. About 120 deer are killed each year in collisions with automobiles.

North of Boulder Mountain Park, the open space system protects highly developable foothills and benches along an arc which swings east toward the IBM facility near Niwot. Farm silos stand near the massive office and indus-trial park. East of the city, in the lower Boulder Creek area, the open space holdings are extensive but more patchy. Some residents have complained that the withdrawal of 34,000 acres from the tax rolls is fiscal suicide. City plan-ners consider this a groundless argument. Each year, the city spends about $75 per acre to maintain open space and $44,000 per acre if full city services have to be provided—which is more than residential development raises in tax revenue.

The City of Boulder's Open Space Program relies mostly on direct pur-chases of land. In 1989, voters approved a fifteen-year renewal of the sales tax for land conservation. However, even when the money is available, there are times when this approach can be circuitous and frustrating.

BOULDER CREEK

City planners had long wished to create a continuous corridor of open space along Boulder Creek. This surprisingly clear stream flows through the city with the university district on its south bank and downtown on its north. In 1982, the Flatiron Gravel Company, which owned three miles of Boulder Creek frontage, was under the gun to reclaim their creeksite gravel pits. The newspapers carried stories alleging that a Flatiron vice president was in-volved in a bribery scheme. It was a public relations disaster for the company.

At the same time, Don Walker of the Boulder-based Colorado Open Lands (COL) gave a speech at a meeting of the Colorado Statewide Gravel Producers Association. Flatiron executives took notice, and Walker and city officials began negotiating with the company. The conservationists wanted to buy Flatiron's land to create a mosaic of bikeways, riparian habitats, and developed parks which would provide flood control and a linkage with the projected Lower Pearl Street Beltway. The gravel company wanted money and a way to improve its public image problems.

Negotiations were torturous. Finally, in 1985, a solution was reached. Flatiron Gravel donated its 81 acres to COL and restored the land to something close to its original condition. COL then sold the land, which was worth $11,750,000, to the City of Boulder for $350,000.

In the former gravel pit, a below-water-level concrete structure with clear plexiglass windows was built along Boulder Creek as a fish observatory. The popular trout-watching facility was a joint effort of the city, Trout Unlimited, and the Boulder Flycasters. The Flatiron Gravel Company even donated the cement. Catch and release regulations are now in place to protect the fishery.

Flatiron emerged with vastly improved public favor, a bothersome reclamation problem solved, and a sizeable tax break for donating the land. The property had been zoned for industrial use, and the company could have easily sold it at a corresponding value. COL received $350,000 to cover project costs and to add to their statewide endowment fund. Boulder County acquired a highly coveted linear park at a fraction of its fair market value. The project stands as a model of how conservation projects can benefit all concerned. Colorado Open Lands is the most successful trust in Colorado. Over 30,000 acres have been directly protected by the group, with another 20,000 acres conserved by projects where COL provided planning and technical expertise.

WHITE ROCKS

Downstream on Boulder Creek lives a woman with unusual ranching credentials. In 1960, a Wellesley College graduate named Ricky Weiser came West and bought land along Boulder Creek, east of the city, and began to raise cattle and horses. Over the years, subdivisions began to encircle her property. In the mid-1980s, Weiser initiated a complex set of negotiations with the City of Boulder and The Nature Conservancy aimed at protecting her ranch.

Weiser's land encompassed the so-called White Rocks area where Cretaceous Laramie and Fox Hill sandstones display rectangular jointing and turtleback surface mounds. A rare fern, *Asplenium andrewsii*, was found growing in the porous sandstones along groundwater seepage zones. The only other locales for this fern were Flagstaff Mountain in Arizona and in the Mexican state of Chihuahua. It was feared that if the property were sub-divided, domestic wells would lower the water table and kill the ferns.

In the end, most of Weiser's 1,000-acre property was conserved by a bushel of land-saving techniques. A conservation easement was granted to The Nature Conservancy on 200 acres harboring the fern colony. The 105-acre White Rocks Natural Area was created. The city purchased some land, and the county holds development rights on other parcels. Weiser retained the right to build up to seven houses on some of the least sensitive land.

In 1988, Ricky Weiser gave a tour of the White Rocks portion of the con-served area. She is an intelligent, no-nonsense person—imagine Katharine Hepburn tending Herefords. She strode through the grasslands wearing a wide-brimmed straw hat, spouting the Latin names of plants and telling stories about herons, kingfishers, and beavers.

BOULDER COUNTY OPEN SPACE DEPARTMENT

The primary goals of the Boulder County Parks and Open Space Depart-ment are "the protection of land as buffers between communities and the conservation of river and creek corridors." The county program has pro-tected over 16,000 acres and manages those lands for recreation.

Residents of the Boulder area are among the most responsive in the county. However, people in the rapidly growing communities of Louisville, Lafayette, and Broomfield seem less concerned. An Open Space staff member believes that Boulder and much of Boulder County support conservation work, while "Louisville and other cities want to fill all the open land with houses and zone everything as prime residential districts."

Boulder County has assembled an exceptional system of properties throughout the mountains, foothills, and plains. The 1986 Boulder County Comprehensive Plan contained maps of vital elk range and migration corri-dors, critical plant associations, rare plant sites, high quality riparian habitats, and key open space. The plan explains that "open space may be expansive, encompassing several hundred acres to buffer urban development or it may be as small as a trail easement or common area adjacent to rural residential development." Open Space personnel note that "while the county has pro-posed that certain areas be protected, implementation is based on the coop-eration of private landowners and our ability to pay." Boulder County plan-ners concur that negotiation is the mainstay of their entire system.

Boulder: A Model Open Space Community

MOUNTAIN PRESERVES

Boulder County's most spectacular open space sites are the mountain preserves known as Walker Ranch, the Betasso Preserve, and Rabbit Mountain Park. The mountainous 2,566-acre Walker Ranch is the largest parcel in the county open space network. Lying southwest of the city in the South Boulder Creek drainage, the ranch contains Douglas fir/ponderosa pine forests and open meadows which are popular spots for hiking and picnicking.

In 1856, James Walker filed a homestead claim in the area. Over the years, Walker was paid by an English mining syndicate, which worked the ranch's quartz lodes along the andesitic Langridge Dike for gold. With this capital, Walker assembled a 6,000-acre ranch.

In 1976, Boulder County wished to protect the area from suburbanization and entered into a lease-purchase agreement for 2,556 acres of the property at a cost of $2.5 million. The mountain park ranges in elevation from 6,700 to 8,200 feet and contains mule deer, black bear, wild turkey, and coyotes and provides winter range for elk. Due to the ranch's cultural importance, most of it has been listed on the National Register of Historic Places—the largest such site in Colorado. Boulder County also leases 1,132 adjacent acres from the Bureau of Land Management, which effectively increases the total park area to 3,698 acres.

The 773-acre Betasso Preserve extends over the north-facing side of Boulder Creek Canyon, west of the city. In 1912, the Blanchard family homesteaded 160 acres in the canyon bottom. Three years later, a hard-rock miner named Steve Betasso bought the parcel. Betasso and his sons, Dick and Ernie, extracted a decent living from gold and tungsten ores in Tertiary replacement deposits in the Precambrian Boulder Creek granodiorite. Using their mining earnings, the family bought more ranchland. By the 1970s, the pressure to sell out for development was extreme. However, Ernie Betasso, who still lived on the ranch, chose instead to sell the lion's share of the property to Boulder County as a preserve for $1 million. Today, Bummers Rock, a dome of granodiorite, is a popular destination of day hikers.

The scenic Rabbit Mountain property lies 5 miles west of Longmont at the edge of the Front Range. It is noted for its classic hogbacks of Dakota sandstone and Benton shale which rise above the flat-lying Pierre shale of the Great Plains. In 1864, Columbus Weese settled at the base of Rabbit Mountain to ranch. His daughter married Jack Moomaw, a forest ranger with a strong land ethic. In 1984, the Moomaws' granddaughter, who had inherited the ranch, sold the property to Boulder County.

Colorado: The Icon

The preserve contains a splendid ponderosa pine/mountain mahogany/ blue grama grass ecosystem. Prickly pear cactus are common. Miner bees, which tunnel into the property's soft sandstone to lay their eggs, depend on prickly pear pollen for their primary food. Raptors thrive by hunting in the preserve for ground squirrels and prairie dogs.

SUBDIVIDING AND CONSERVING

There are reasons for optimism about the future of Boulder County's program. At present, there is a staff of twelve, and some $1.4 million is spent each year on land acquisitions. Yet one official believes, "If we were to propose increasing the sales tax or real estate transfer tax, people would certainly object."

The subdivision review process is even used by the Boulder County Planning Department as a vehicle to conserve land. The 1988 county subdivision regulations mandate that "any subdivided land must be reviewed prior to the issuance of a building permit." However, the Non-Urban Planned Unit Development (PUD) section of the regulations are having the most impact on land protection.

In 1986, a large area of the county was "downzoned" from commercial and residential uses to an agricultural classification allowing only one home per 35 acres. However, a "density bonus" transfer system was put in place which allows two homes under specific conditions. The two structures must be clustered on no more than 25 percent of the parcel, with the remainder of the land categorized as a "preserved area" which is not eligible for future subdivision. A conservation easement is then granted to the county over the open land for a set number of years.

These non-tax-deductible, short term easements are actually only non-binding covenants. They formalize an agreement between the landowner and the county that the property will not be further developed in the immediate future. According to the 1988 Boulder County subdivision regulations, "The period of assurance of non-development of the preserved area is not time specific, but is rather determined by and tied to changes in the applicable land use regulations, annexation, or inclusion of a Community Service Area." According to a planning official, "Landowners can later apply to build an additional two or three units on each 35-acre parcel."

The system arose as a political compromise. It is used as a holding action with the hope that large properties will be developed using clustered housing with perpetual conservation easements used to preserve the surrounding open space. Critics argue that participating landowners are merely reducing their property taxes while temporarily holding off on subdivision plans. However, as long as the land is zoned for agriculture there is a chance the open space will remain.

THE NATURE CONSERVANCY IN BOULDER

The Colorado Field Office of The Nature Conservancy (TNC) is located in downtown Boulder, just off the Pearl Street Pedestrian Mall. TNC has protected some 70,000 acres in Colorado, spread over 61 sites. In Boulder County, TNC has secured 90 acres of alpine meadows near Caribou, accepted a gift of 12.5 acres of remnant short grass prairie near Broomfield, and received a conservation easement on the 183-acre Rangeview Ranch which contains riparian habitat and elk range along the South St. Vrain River near Ward. TNC also purchased one cubic foot per second of water in Boulder Creek to help maintain instream flows and protected a rare fern on Ricky Weiser's land at White Rocks. The Conservancy is currently negotiating for a 320-acre conservation easement on the Sheep Mountain Ranch along the North St. Vrain River. Potential new projects abound.

YET MORE LAND TRUSTS

The Wilderness Land Trust is the invention of former Aspen attorney Jon Mulford. The mission of this Boulder-based trust is to secure the protection of privately owned inholdings within Federal wilderness areas. Nationwide, there are 452,000 acres of such inholdings, 13,000 acres in the U.S. Forest Service wilderness areas of Colorado. Not all of these lands are threatened. Due to a convoluted legal quirk, about one-half of the national total consists of lakebeds owned by the State of Minnesota in the Boundary Waters reserve. However, considerable additional acreage consists of lakebeds in Alaska, which Governor "Wally-World" Hickel wants to open to placer mining for gold.

Of the 13,000 acres of inholdings in Colorado, 3,050 are owned by the City of Durango within the Weminuche Wilderness. These sensitive high-elevation San Juan Mountain habitats are unlikely to be jeopardized in the future, since they can only be used as municipal watershed. The remaining 10,000 acres, consisting largely of former patented mining claims, are vulnerable to recreational homesite development. These parcels are being prioritized by the Wilderness Land Trust for protection. Lands near streams and road access are the most at risk.

The trust operates by accepting gifts of land or purchasing tracts and then transferring them to the U.S. Forest Service for wilderness management. Thus far the following projects have been completed: 38 acres in the Hunter Fryingpan Wilderness near Aspen, 37 acres in the Mount Massive Wilderness and 75 acres in the Holy Cross Wilderness areas near Leadville, and 160 acres in the Indian Peaks Wilderness above Boulder. The Wilderness Land Trust began as a Colorado group, but its future emphasis will be on protecting inholdings nationwide.

Boulder is also home to two other groups: The Boulder County Nature Association (184 acres conserved) and the Boulder County Land Trust, which is targeting large ranches for conservation easements.

CONCLUSION

There is a running race held every year in the city called the "Bolder Boulder." Bolder is certainly an appropriate word for the place. The diversity of successful public and private land conservation organizations working in this unique setting exceeds that of any other jurisdiction in the Rocky Mountain West. Local government open space programs (which function as land trusts) and private land trusts have conserved over 50,000 acres. Despite occasional minor squabbles over turf, it appears that there is a remarkable level of communication and cooperation. Given that the population of the county grew from 132,000 in 1970 to 226,000 in 1990, this ability to work together is not only critical, but, like everything else in Boulder, highly unusual.

Boulder represents a progressive exception to the conservative ways of the West. What has happened in this "green" college town may not be repeatable in most places, but communities all over the region would benefit by studying Boulder as a model of open space stewardship. Perhaps some of the optimism will prove contagious.

8

SKI COUNTRY:
UNCONTROLLED DEVELOPMENT
AND INDUSTRIAL TOURISM

The Colorado Mineral Belt slants diagonally across the mountainous portion of the state like the tail of a comet. Various theories on megatectonics have evolved to explain its southwest to northeast orientation. However this alignment arose, its presence accounts for the location of nearly every town in the state's high country. Telluride, Silverton, Creede, Crested Butte, Aspen, Climax, Leadville, and Georgetown are a few of the settlements born of mineralization. These towns of the Belt archipelago boomed from the 1860s to the 1890s and mostly went thud after that.

SILVERTON

From a distance, Silverton seems more like a daguerreotype image than a real community. For a brief period during the 1880s, Silverton claimed 5,000 residents and was the hub for the surrounding silver camps at Rico, Ophir, Ouray, and Telluride. The collapse of silver prices in 1893 left the town eerily quiet when miners bailed out to gold camps. For decades, the hangers-on in Silverton wallowed in economic depression with only the self-deceiving hope of a mineral revival and shots of whiskey to keep their spirits warm against the high, cold air.

The Durango and Silverton narrow gauge railroad began operations in 1882. By the 1940s, summer train service began, bringing camera-laden tour-

^^^^^

ists up the Animas Canyon to Silverton. The decaying buildings, grey snow-capped peaks, and deep blue skies sold a lot of film.

Today, increasing numbers of tourists pay hefty prices to ride in refurbished orange railcars to Silverton for an afternoon. Whenever a train arrives at the wood frame depot, tourists fan out to browse in clothing stores, rock shops, pottery stands, and art galleries. Others head straight to the Fudge Factory, Swanees Healthy Junk Food, or the Hamburger Emporium. Six "Kodak Film Sold Here" signs hang on one block. The King Frog 45-Minute Photo Lab operates next to the gussied up false front of Natalia's 1912 Restaurant. Visitors crane their cameras at odd angles to avoid including the many trailer houses, propane tanks, satellite dishes, snowmobiles, and scrap heaps in their carefully framed images.

The community seems to be having a debate over how health conscious they want to be. A Swiss-style stone hotel has "For Non-Smokers" in block letters on its side. The downtown Wyman Hotel has "A Luxury Hotel for Non-Smokers" written on a red canvas awning over its front door. In rebuttal, the Bakers Park Rendezvous Family Restaurant has a sign on its door which reads, "NO Provisions Made for Non-Smokers!"

Although some mining and miners remain, Silverton is unmistakably a mountain tourist town. And tourism, for good or ill, is what the 716 residents of this once raw-boned hellhole depend on for their survival.

The 14,000-foot peaks of the San Juan Mountains encircle Silverton. There are no foothills. The mountains simply rise straight up from the upper valley of the Animas River. North of town on a steep mountainside is a fifteen-foot-tall granite statue of Jesus placed in a shrine of red and grey stones. There are several of these "Christ of the Mines" statues in the Rockies. The one in Silverton, built in the 1950s as a statement of faith that God would revive the community's dire financial condition, was a kind of divine economic development strategy.

In a curious way, it worked—not with a silver boom, but with El Niño (the Christ child), the shifting Pacific Ocean current which periodically brings wet years to the mountains. As air masses rise over the Continental Divide, their moisture fast-freezes into the powdery snow which incites skiers to ascend the mountains of Colorado like pilgrims on Mount Fuji. The prayers of people in Silverton and other moping mining towns of the region were answered by orography, not ore.

Silverton benefits economically from the nearby Ouray, Purgatoire, and

Tamarron ski areas but itself has not become a renowned destination area resort. Places such as Aspen and Telluride have had their ways of life turned cattywampus. Today, the fortunes made in the skiing industry and by life-style realtors in parts of Colorado dwarf even the most rabid miner's fondest metallic fantasy.

SUMMIT COUNTY

The July 1989 issue of *Snow Country* contains an article entitled "Where the Growth Is." It described the nationwide population and development explosion surrounding America's major ski resorts. Summit County, home to the Keystone, Arapahoe, Copper Mountain, and Breckenridge ski areas, is described as a "super-dynamic county." Only Summit County, Utah, with development surrounding the Park City, Park West, and Deer Valley resorts, was expanding at a more rapid rate.

The resorts of Summit County, Colorado, tally approximately 2.3 million "skier days" each year, or about 30 percent of the state's total. Much of this is due to I-70, which affords easy access for Denverites to the Gore Range ski slopes. In 1970, Summit County had 2,665 residents. By 1980, it had grown to 8,847, and by 1990 to 12,881. In addition to this 383 percent growth in 20 years, the Summit County Master Plan estimates that 26,000 people own dwellings or timeshare condos in the county. All of this development is compressed into a narrow strip of private land in the Blue River Valley.

North of Dillon and Silverthorne, the valley still contains significant amounts of ranchland. Cows and calves graze in the pastures, sagebrush grasslands, and aspen groves adjacent to recreational developments. In the early 1980s, The Nature Conservancy (TNC), concerned about this trend, began negotiating with landowners to protect the valley's ecological and open space resources.

In 1981, Bruce Bugbee and Associates of Missoula, Montana, prepared an ecological baseline study for TNC on the 3,130-acre Mount Powell Ranch near the Green Mountain Reservoir. The Blue River forms the ranch's east border for three-and-one-half miles. The fishery is among the best in the state, with plentiful brown, brook, and rainbow trout, as well as the prized Kokanee salmon. Six bald eagles winter on the water. Hundreds of mule deer reside or winter on the ranch, which serves as a migration corridor across the valley. The property is also critical winter range for 350 elk—more than any other single ranch in Summit County.

A conservation easement which protected the entire Mount Powell Ranch from subdivision was donated to TNC at the end of 1981. The easement was later transferred to the American Farmland Trust, whose mission is to protect agricultural land. Despite this heartening project, most of the remaining rural land in the county is in zoning classifications which permit development at a habitat destroying density of one dwelling per 20 acres or less.

A new organization, the Continental Divide Land Trust in Breckenridge, is now attempting to become legally incorporated to contend with the intense development pressure throughout Summit County.

EAGLE COUNTY

Eagle County, with the Vail and Beaver Creek resorts, and Routt County, with Steamboat and Howellson, have also been ranked as "super-dynamic" counties. The real estate market around the Vail/Beaver Creek complexes is, at this point, an irresistible force. Vail's prefab Swiss village ambience sprawls over what was pasture land until the 1950s. The ski slopes extend nearly into the hotel parking lots. The setting is claustrophobic. There is a constant roar of semis on I-70, which bisects the development. North of the interstate, houses, townhouses, and condominiums slouch closely together at the base of a hill—classic condo bondage, as some locals call it. New roads wind further upslope to serve the numerous dwellings under construction.

Perhaps nowhere else in the Rockies are such poor building lots so expensive. People are paying top dollar to watch, hear, and smell Interstate 70. In the valley bottom, housing and commercial development extend from Vail more than 10 miles westward to Avon. Many people with jobs in the Vail/Beaver Creek area cannot afford to live there. Some commute from Eagle, which lies 30 miles west, where large trailer parks are set out in the sagebrush and Gambel oak.

The Eagle County Land Conservancy was formed in 1982 in response to the rampant conversion of ecologically important agricultural land to recreational housing. It's been rough work. The land trust thus far holds title to only 60 acres and has received no conservation easements. The group even has trouble raising the $1,500 needed to pay its annual property tax bill.

This is an extraordinary amount of tax. Local assessors apparently are categorizing open land as residential property. This tends to cripple agricultural operations and forces ranchers to sell out. A member of the Eagle County Land Conservancy is not optimistic: "Eagle County people aren't thinking about donating land or easements, they want money—BIG money!" For a land trust with a full-time staff, a setting like Vail could be advantageous. There might be excellent fundraising opportunities among wealthy landowners and numerous chances to negotiate partial development/partial conservation easement designs for key parcels. However, even The Nature Conservancy has had problems getting much done around Vail. Its sole project has been a donated 3.5-acre parcel of deer migration habitat along Gore Creek, which TNC then transferred to the Colorado Division of Wildlife. The notorious transience of Vail's population seems to work against long-term stewardship taking root.

ASPEN AND PITKIN COUNTY

The aspen (*Populus tremuloides*)—not the blue spruce—is the real state tree of Colorado. The aspen's white to light olive bark, rustling leaves, and fall dance of color are the signature of the high country. To landscapers and realtors, the aspen is a stylish sort of tree. It defines the essence of mountain decor.

Pilgrims travel hundreds of miles each autumn to see the aspens' yellow and gold leaves against the blue sky above the Maroon Bells Range near the city of Aspen. This small resort community of 5,000 is known worldwide as a place where the wealthy gather to play and spend in style. *Populus tremuloides* are everywhere. They line the ski runs, border golf course fairways, and frame views from picture windows. The species is a kind of fashion accessory and Chamber of Commerce real estate totem all at once. This hasn't always been so. In the past, this tree was seen as just a weed which covered up ore deposits.

In 1873, Ferdinand Hayden mapped the Roaring Fork Valley and noted its geologic similarity to Leadville, where rich silver strikes had been made. By 1879, four prospectors entered the valley and saw promise in the rocks of Aspen and Red mountains. In honor of encroaching on the Ute Reservation, early arrivals from Leadville (then the second largest community in Colorado), called the initial settlement Ute City.

The Aspen Town and Land Company was formed in 1880, establishing the first of what would be scores of real estate offices in the valley. Veins within the mountains proved to be full of silver. One 2,054-pound "nugget" was assayed as 93 percent pure. By 1883, 800 people swarmed over the growing town. The local newspaper, *The Rocky Mountain Sun*, proclaimed in 1887 that "Aspen today is the most prosperous city in Colorado" (in Rohrbough, 23). The Denver and Rio Grande Railroad reached town as production from the Mollie Green and other mines totaled $8 million per year.

The city townsite was platted at the base of the mountains along the banks of the icy Roaring Fork River. Grand homes, restaurants, and stores filled the preferred portions of the grid, while the small combustible clapboard shacks and boardinghouses of the miners lay on the periphery. The 1890 census tallied 5,108 people in Aspen; all of Pitkin County, named for Governor Frederick Pitkin, totaled 8,929 residents.

Although times were good, the price of silver began to drop. This prompted Aspenites to form the Silver Club, whose original credo might be summed up as "mine more and more silver and sell it at higher and higher prices." At the same time, nearby Leadville, long the second largest producer of silver in the world (behind only the Mexican mines at Santa Barbara), started to feel the strain. By 1892, silver demonetization was seen as the likely outcome of an untenable U.S. policy of gold/silver bimetalism.

Several western European countries began moving toward the gold standard. In 1893, a revolution in Argentina hurt several British companies just as England closed its Indian silver mines and began to reduce its investment in American companies. In the process, several large British-financed East Coast corporations which had backed Colorado mining companies failed.

National financial panic ensued. Even currency backed by gold weakened. President Grover Cleveland had long believed any decline in the strength of gold could be linked to the government's subsidy program enacted under the Sherman Silver Purchase Act of 1890. In October of 1893, the Sherman Act was repealed. The news hit Colorado like an iceberg below the waterline. Within 30 days, nearly every silver mine in the state shut down.

In one week, the number of miners in Aspen fell from 2,250 to 150. *The Rocky Mountain Sun* moaned: "The Crisis at Hand—The dreaded emergency is now upon us . . . a situation as deplorable as the imagination has ever conceived" (in Rohrbough, 219). A few mines continued to operate at greatly reduced capacities, but the silver-gilded era was history.

The valley remained economically depressed from the 1890s to the 1950s, when a nascent ski industry and second home market began to expand. Between 1950 and 1960, the population of Pitkin County climbed 45 percent. During the 1960s, the county grew 160 percent as the local ski slopes attracted international attention. The 1970s and 1980s produced undulating but strong expansion as Aspen became a coveted address. By 1990, the county had 12,661 people.

Aspen is built on a glacial outwash terrace in the narrow valley of the Roaring Fork River. Near the city, this northwesterly flowing river is fed by Castle Creek, Hunter Creek, and Maroon Creek (whose headwaters are in the much-photographed Maroon Bells Range). Downvalley in the watershed, Woody and Snowmass creeks join the flow. The Fryingpan River enters near the town of Basalt, and the Crystal River enters from the southwest at Carbondale. Finally, the Roaring Fork debouches into the Colorado River at the railroad resort town of Glenwood Springs. The high peaks of the Elk Mountains and Sawatch Range define the valley's skyline. Lower foothills and benches serve as prime big game winter range habitats. However, the physical geography of the county, while handsome, is not what makes it unique.

ROCKY MOUNTAIN HIGHS

In Aspen, everyone seems very well tended. The place carries the smell of serious money, like many insular resort communities. Today in Aspen that scent is as strong as patchouli was in the '60s. Downtown is full of shops with names like "The Best of All Worlds" and "Les Chefs D'Aspen—Gourmet Take Out." Clothing galleries crowd every block. A modest lunch can easily run fifteen dollars for petite portions of nouvelle cuisine. In the center of town is the Hyman Avenue Pedestrian Mall, a smaller but vastly upscale version of Boulder's Pearl Street. The city's Saab police cars cruise discreetly amid the crowds. Recreational shopping seems to fill the days between the end of ski season and the summer music and art festivals. The local economy generates nearly a quarter of a billion dollars in retail sales annually.

The ski runs of Aspen Mountain drop directly into town, where a new glass and steel chairlift is under construction. A full day lift ticket here costs $45—the highest in Colorado. Skiing is also available at nearby Aspen Highlands, Snowmass, and Buttermilk. Over 1.4 million skier days are logged each year in Pitkin County, which, perhaps not coincidentally, also has the greatest density of lawyers in the state outside of Denver. World Cup ski races and ski exhibitions by Hollywood headliners to benefit charity are part of life in

Aspen. Aspen is a de rigueur stop for the international jet set and a seasonal home to Jack Nicholson, Martina Navratilova, Steve Martin, Don Johnson, Melanie Griffith, and scores of other celebrities. While the pace is slow relative to L.A., Aspen is to the Rockies what Singapore is to Malaysia.

Aspen is a place of disparate voices. Two of these, journalist Hunter S. Thompson and singer John Denver, are not only notable but, at first glance, antithetical to each other. The Thompson image arose from his books such as *Fear and Loathing in Las Vegas*. His image is gonzo, debauched, and angry. Denver's media persona is the fresh-faced mountain folksinger—clear-headed and clean-living. As usual with such creations, the truth is more complicated.

I met Hunter Thompson during a consulting trip to Aspen. The Aspen airport was closed for repairs, and we landed in Rifle, the then still-hopeful oil shale town. I had reserved a rental car, but Thompson apparently hadn't and moped in a corner of the Rifle airport, smoking a cigarette in a long holder. I offered to drive him home. As we drove past Glenwood Springs, he insisted we stop at a liquor store. He emerged very pleased with himself, got in the car, and pulled a large porcelain Elvis whiskey decanter out of a paper bag. It had a music box in its base which played, "I Can't Help Falling in Love with You."

When we reached his lair, Owl Farm, he invited me in for dinner. Over drinks, we talked about football, Nixon, and eventually conservation easements. Thompson expressed interest, so the next day, we walked his 110-acre property, which turned out to be excellent deer and elk winter range. He would later conclude that it just wasn't the right time to go ahead with an easement.

Hunter Thompson is extremely bright, well read, and beyond the Gonzo, more than a little shy. He once ran for sheriff of Aspen and barely lost. Although he often does things tongue-in-cheek, he cares deeply about what is just. He also has a short fuse when it comes to land developing "greed-heads." The daily nonsense of the world hits him hard, and the crazed on-the-mark political journalism he produces seems as much truthful comic relief as frontal assault.

On a recent trip to Aspen, some people of the Woody Creek area expressed annoyance at Thompson's surly behavior. One local even volunteered, "Hunter's an asshole." However, a photo of him still hangs in the Woody Creek Tavern. Thompson's house is bedecked with a poster which proclaims, "Meese is a pig." He is a life member of the NRA. His garage door is adorned with something that looks like a swastika. It turns out to be an ancient East Indian symbol, the mirror image of the Nazi emblem. Thompson is a man of epic contradictions.

Then there's John Denver. It's a fair bet that Governor James Denver of the Kansas Territory, which included parts of what is now Colorado, never expected to be immortalized by both a major city and a blond, bespectacled folkie. The songs of John Denver have crystallized Colorado in the national consciousness to a degree only dreamed of by Chambers of Commerce. In the early 1970s, Denver's song "Rocky Mountain High" played a central role in sanctioning the state as a must-see attraction.

John Denver didn't start the 1970s boom in Colorado, but he was an unwitting promoter of the state's realty. In that era, if you told people in Boston or L.A. you lived in the Rockies, they invariably said, "Oh, Colorado sounds so nice!" Denver's television specials featured the soaring mountain scenery around Aspen, which he has called home for years. The Maroon Bells, Aspen Mountain, Mount Sopris, and other local landmarks were beamed in color across the nation and eventually around the world.

Today, in his private plane, Denver flies in and out of Aspen to concerts and increasingly to environmental conferences. He has formed the Windstar Foundation to promote ecological quality, sustainable ways of life, and world peace. According to its pamphlet, the purpose of Windstar is to:

inspire and bring forth ideas and models that contribute to a world that works, through research, demonstration, and education. Our mission is to provide ideas and inspiration for responsible, informed action.

The Foundation conducts an annual "Choices for the Future Symposium" ($975 per couple), presents the Windstar Award (recipients have included Jacques Cousteau), and markets an eclectic mix of audio tapes with titles such as "Human Choice and Human Responsibility," "Wholeness," "Conflict Resolution," "Grassroots Organizing," "Children with AIDS," "Hunger," and "Space." Books sold by the Foundation's mail order business tend to have New Age titles such as *Hug a Tree*, *The Magic of Conflict*, and *Your Edible Landscape Naturally*. Members include environmentalists, naturopaths, ex-astronauts, spiritual seekers, ex-cowboy actors, organic gardeners, and New Age entrepreneurs. In 1988, the group began the Windstar Connection program designed to bring together people in many cities to discuss the big issues of the day. Windstar is an idealistic organization built on the trust that the "seeds" planted today will grow into a better world.

The Windstar Foundation headquarters is located up the old Snowmass Road in lush pastureland below the stunning crestline of Snowmass-Maroon Bells Wilderness. The entrance to the visitors' area displays a large photo of a sign which says "Leave your ego with your shoes." This main building is a new solar-heated structure with an interior greenhouse, conference rooms, offices, and hallways cluttered with boxes of literature. The central meeting room has a large wooden sculpture of the flukes of a diving whale. A stereo sits in the corner next to a TV and VCR. The album collection ranges from Creedence Clearwater Revival to Joan Baez. Staffers are friendly and busy answering phones and doing errands. The greenhouse, bursting with flowers, is bordered by pools full of carp and goldfish.

Out in the adjacent field is the most striking feature at Windstar—a silver-white "Biodome" 50 feet in diameter. The solar-heated dome is designed to demonstrate that food can be produced year-round in Colorado's tough climate. However, the agricultural production from the Biodome and nearby outdoor gardens is not particularly impressive. A hardworking fifteen-year-old with a small plot, a hoe, some manure, seeds, and a hose could easily produce twice the food grown at Windstar. Nonetheless, the Foundation's attempt to link organic agriculture, social consciousness, and peace activism is well-intentioned and does some good.

Hunter S. Thompson and John Denver have little in common except a substantial concern for the future. Given the rapid population growth and quickening pace of life in the Aspen area, the landscape can use all the supporters it can get. In this regard, Thompson and Denver share far more than either might care to acknowledge.

PLANNING AND CONSERVATION IN ASPEN

The development pressure on open lands near Aspen is intense and escalating. Land-use planning began to contend with this growth in 1966 with the passage of the Aspen Area General Plan. By 1974, Pitkin County was completely zoned. A Growth Management Plan (GMP) was devised in 1977 and a Growth Management Quota System (GMQS) adopted in 1983. Development proceeded undeterred.

In 1986, a "Down Valley" comprehensive planning process was initiated. The Down Valley area consists of rural areas outside Aspen, including Snowmass Village, Woody Creek, and the valley lands north along Route 82 to Basalt. The principal objective of the plan has been to identify the highest priority land for preservation and to suggest conservation strategies which also provide suitable locations for new development.

The Down Valley area consists of 74,000 acres, of which 69 percent is private land. It supports a population of 3,280 people and contains big game migration corridors, summer range, elk calving areas, riparian ecosystems, and scenic agricultural lands. Eighty percent of the private land is owned by only 70 people.

For years, Pitkin County planners relied on regulatory systems to wrestle with changes in land use. Unique concerns arose. The planners shied away from any system which anticipated growth. To them, it was safer to react, since showing preferred growth centers on maps might provide incentive for development to occur. It was feared the county's active land speculators would buy up growth center lands and jack up prices beyond the reach of working class residents.

Under the Growth Management Quota System (GMQS), each proposed development was reviewed and awarded points on whether it provided for "employee housing," reduced density below that allowed by zoning, or met other community goals. The GMQS was seen by the public as too complex, since lawyers and consultants were needed to get a subdivision approved. The process involved considerable expense, which acted as an incentive for landowners to split up large blocks of land rather than create just one or two tracts. The large developers often played the system to their advantage by securing points needed for passage by opting for larger lots rather than by making allowances for affordable housing units. As a result, fewer but larger lots were surveyed. These estate-sized tracts were immediately attractive to wealthy newcomers, and the rural landscape became even more gentrified.

Due in part to the regulatory maze erected by planners, more land also began to be subdivided under the state's 35-acre exemption law. For several years, Pitkin County chose to review these large supposedly unreviewable tracts under their GMQS, until they were successfully challenged in court. Planners indicate that today, those same 35-acre splits are still reviewed as general subdivision submissions. Access, water supply, and utility services are checked prior to issuance of a building permit.

The Down Valley planning process stirred up the slumbering residents of rural Pitkin County. The planning staff took aim at protecting agricultural land whose main "crops" were open space, wildlife, and clean air and water. The plan stated: "The challenge facing Pitkin County is how to reward owners of agricultural lands for the production of 'crops' for which the marketplace provides no compensation." In January 1986, *The Aspen Times* ran a story entitled, "Land Trust Concept May Offer Compensation for Ranchers." Aspen was already using local sales tax money for some open space planning, but what was being discussed was far more grand.

As part of the comprehensive plan, prime agricultural lands, irrigated fields, critical wildlife habitats, and scenic resources were mapped. Three implementation systems were studied: agricultural zoning, the transfer of development rights (TDR), and "land use compensation programs." The planners concluded:

> *The most successful land preservation programs in the nation have some level of compensation to landowners. These programs which feature the acquisition of scenic or agricultural easements are far more desirable than fee simple purchases. TDR programs . . . are generally unsuccessful or are very difficult to implement . . . Agricultural zoning alone in the absence of land use regulatory and compensation programs may not ensure the continuation of agriculture.*

In 1986, representatives of the American Farmland Trust (AFT) were called in. In Aspen, Doug Horne, Will Shafroth, and Will Grossi met with local planners and explained how AFT operates nationally to conserve agricultural land. The county decided that TDR and purchase of development rights (PDR) programs should be instituted.

The public became confused and angry at trying to sort out GMP, GMQS, TDR, PDR, AFT, and other acronyms. One Woody Creek rancher cautioned, "We're suspicious. We have gotten beat up every time on this. The planners ought to get out of their rooms and get on the land!" Of course, he was right. Another rancher grew dramatic: "Ranchers are faced with a decision; either build or die. You're trying to stop growth and have your own little domain here. And you want to keep it unpopulated." Despite Aspen's progressive image, the sovereignty of county government weakens dramatically in rural areas.

The Down Valley Comprehensive Plan was finally passed in January 1987. Various sources of money to implement the land conservation program were discussed. These included: additional local sales tax, property taxes, lottery funds, park dedication fees, general funds, and the existing real estate transfer tax. Escalating rural resentment made such talk appear premature. One local resident barked, "I want freedom! My retirement plan is tied up in my land." Another revealed, "We don't feel we're partners in the solution. We feel we're being dealt with as the problem." A prominent rancher simplified the clutter by spelling out the bottom line. "Open space is great," he said, "but you have to buy it." In this case, rural residents unknowingly directed planners to seek out an approach already in use.

Colorado's oldest land trust, The Aspen Valley Land Trust, was formed in 1966 with the goal of "preserving and enhancing natural beauty and providing trails and other recreational facilities" throughout the Roaring Fork Valley. Trust members realized long ago that with 80 percent of the county in public ownership, all land development pressure would focus on a small, ecologically fragile private land base.

It is surprising that during all the discussions about land conservation during the completion of the Down Valley Comprehensive Plan, The Aspen Valley Land Trust was not highlighted. In the 1960s, the then Pitkin County Parks Association aided in the establishment of four parks. The 1970s saw the passage of a 1 percent sales tax for open space acquisition. The trust was actively involved in establishing this land fund.

Using these funds, Pitkin County secured land for the Aspen Golf Course, bought the 175-acre North Star Ranch as a big game and waterfowl habitat (with help from The Nature Conservancy), and built the Rio Grande and Hunter Creek hiking trails. During this same period, The Nature Conservancy bought 170 Down Valley acres which were later transferred to the White River National Forest and received a 156-acre gift in the popular Hunter Creek drainage which also ended up in U.S. Forest Service ownership. In the middle 1970s, under the urging of The Aspen Valley Land Trust, Aspen established the Open Space Advisory Board, hired an open space coordinator, and adopted the "Aspen Area Open Space Master Plan."

In the 1980s, the local government began to unravel as conservative politics swept into Aspen. The open space sales tax was converted to a general slush fund and used on projects such as remodeling City Hall, reconstruction of the Wheeler Opera House, and a $500,000 diversion to the city's management and administrative bureaucracy. In addition, the Open Space Advisory Board was abolished and its coordinator fired. The city and county are today only minor players in the land conservation arena. Despite the passage of the

Down Valley plan, conservation matters are essentially left to private land trusts.

But here the story grows confusing. A May 1989 exit poll in Aspen revealed that the public is overwhelmingly concerned about escalating land development. Twenty-one percent of those polled wanted zero growth, 54 percent wanted less growth, 18 percent favored the present pace, and 5 percent wanted more growth. Only 1 percent favored complete sack and pillage.

The Aspen Valley Land Trust's track record is enviable. Its current holdings represent a value in excess of $1.5 million and consist of 8 conservation easements totaling 575 acres (including one on former critic George Stranahan's Carbondale Ranch), ownership of 11 parcels totaling 125 acres, and development of an extensive collection of recreational facilities. Nevertheless, after 20 years of trying to work with a seemingly supportive local population, trust members express frustration with the area. With all the sprawling mansions and mega-wealth in the vicinity, the Aspen Valley Land Trust must carefully exact a percentage of each open space transaction to stay solvent. General fundraising is a bust among the local area's many millionaires. It seems they come to Aspen to escape responsibility, rather than to assume it. Local people describe an "Aspen malaise" which influences social responsibility and land-use planning. One conservationist describes the symptoms:

> *Aspenites don't think they need to go outside. "The world comes to us" is the prevailing attitude. But Aspen and Pitkin County are way behind other parts of Colorado in planning and open space protection.*

Pitkin County recently gained a second land trust. In 1988, the Roaring Fork Land Conservancy was formed to conserve rural lands. The trust will focus its efforts on ranches and farms and leave Aspen recreation issues to The Aspen Valley Land Trust. Members believe The Aspen Valley Land Trust is seen as "too Aspen" for the tastes of agricultural landowners. However, the Roaring Fork Land Conservancy has not completed any projects and is losing momentum in the face of the arduous realities of fundraising and project negotiations. Another land trust, the Aspen Center for Environmental Studies, has conserved 475 acres of wetlands and other wildlife habitat while mostly engaging in programs which promote environmental education. A new group called the Aspen Historic Preservation Trust is in the process of organizing.

CONCLUSION

Over the past 30 years, Aspen has had several incarnations. In *Songs of the Doomed*, Hunter Thompson wrote:

> *What was once the capital of Freak Power has rolled over and is now a slavish service community . . . Where absentee greedheads are taking over the town like a pack of wild dogs . . . Aspen is a big-time tourist town, and only two kinds of people live here—the Users and the Used—and the gap between them gets wider every day. (271–272)*

A recent poll in Aspen revealed that 53 percent of year-round residents believe they will have to move if housing doesn't become more plentiful and cheaper. An extra 2.5 percent has been added to the real estate transfer tax to pay for affordable housing projects. This has come not a moment too soon, since even small suburban houses in Aspen's Snowbunny Addition are out of most people's reach. The Aspen real estate ads take your breath away. A large house on 15 acres sells for $3.5 million, a two-bedroom condo for $590,000, a townhouse for $1 million. A small residential building lot is listed at $750,000. A one-bedroom apartment rents for $700–1,000 a month; a very average three-bedroom house rents for $5,000. An ad in *The Aspen Times* reveals what a major league draw the area has become:

> *Estate Wanted. East Coast family seeks 7,000 square-foot luxury home on 100 + acres between Aspen and Carbondale. $5–15 million. Contact Sotheby's/Colorado Landmark. (May 4, 1989)*

This kind of hyperinflated real estate value works both for and against land trusts. However, for local people who must earn a living, escalation is bad news. Carbondale and Basalt, located north of the county line, are called "employee housing communities." One local woman calls them "servants' quarters." They are towns of trailer houses and station wagons. As Aspen becomes unaffordable and increasing numbers of people move downvalley and commute to work, there will be even more developmental pressure on rural lands which formerly were outside Aspen's zone of influence.

Colorado: The Icon

It is no wonder so many people want to live in Aspen. The surrounding mountains are rapturously beautiful, and the fishing's still good. The ski trails through the aspens form intriguing, even artful, patterned ground. There are lots of good places for relaxing after a day of skiing or hiking. Aspen has survived the classic Western powder progression from blackpowder rifles and miners' dynamite to snow to cocaine to snow again. It has endured a silver panic, destitution, freak power, and international acclaim.

In this decade, the people of Aspen and Pitkin County will face one of their biggest challenges yet—to decide where growth should go and which lands should be conserved in perpetuity. The local governments have recently begun to revive their open space programs. But the job will not be easy.

One local conservationist describes Aspen as "a spiritually bankrupt place, a Sodom and Gomorrah of environmental devastation." When pressed, she explains:

> *People come here and build a massive house ranging from 5,000 to 20,000 square feet which they will use a couple of weeks of the year. Think of the raw and manufactured materials it takes to build a house with a quarter-acre of floor plan! For most of the year, these structures also have to be heated, whether or not the owner is there. Each house also has one or two cars stationed there. These people jet in for a brief stay and imagine themselves roughing it in the pristine West. Trouble is they often have one or two more houses elsewhere in the world, in Hawaii, L.A., or somewhere, which are often just as big. This kind of hyper-construction is appalling, but getting people to see it is an exercise in futility.*

Beyond these general environmental concerns lies the central vexing question which confronts the area's land trusts and planners. How can voluntary land conservation programs better protect the keystone habitats and scenic vistas which define this dazzling landscape? Given the whirlpool of change and emotion in the Roaring Fork Valley and throughout Colorado's ski country, conservationists must continue to look beyond the mountains to the land trust community for inspiration. The 1993 national meeting of the Land Trust Alliance was to be held in Aspen and Snowmass. However, this opportunity was lost when the meeting was moved due to a boycott of the state for its recent anti-gay legislation.

9
SOUTHERN COLORADO: THE OUTBACK

SAN LUIS VALLEY

The San Luis Valley outlasts those who try to claim it. This 50-mile-wide, 100-mile-long, semiarid structural basin is bordered on the east by the Sangre de Cristo Range and on the west by the massive volcanic San Juan Mountains. The southward-flowing Rio Grande bisects the valley on its way toward New Mexico, Texas, and the Gulf of Mexico. The broad intermontane valleys known as North Park, Middle Park, and South Park lie to the north.

The San Luis Valley has been called San Luis Park, the Saguache, and the Rincon Basin. The first settlers in this former Ute territory were Hispanics from Taos who, in 1851, established San Luis as the first town in what is now Colorado. Settlements in present Costilla County, such as San Pablo, San Pedro, and San Francisco spread along Culebra Creek and the San Luis Ditch. This canal, completed in 1852, is still in use and has the first-priority water right of all claims in Colorado. Large Mexican-era land grants such as the Sangre de Cristo (covering most of Costilla County), Baca Number Four (north of Great Sand Dunes National Monument), and the sprawling Maxwell Grant east of the mountains totaled nearly 3 million acres in the San Luis Valley region.

In the 1870s, Anglo land claiming began in earnest. The German Colonization Company established a utopian agricultural community between the Sangres and the Wet Mountains south of present Canon City. Crops failed in the high, cold climate, and most settlers bailed out. In the San Luis Valley itself, the establishment of Fort Massachusetts, which was replaced by Fort Garland, was the first step toward widespread settlement.

∧∧∧∧∧

In 1887, a Mormon missionary named John Morgan led about 80 faithful Deep Southerners into the basin. Later, the Hans Jensen group surveyed a townsite west of the Rio Grande which they named Manassa. Ten-acre allotments were given to each Mormon family. Manassa, Bountiful, Sanford, and other settlements did so well that by the 1880s, some 500 Mormons controlled over 40,000 acres of cultivated fields and pastures. The ditches and canals of Utah served as their model for directing irrigation water from the Alamosa, Conejos, and Rio Grande rivers onto an expanding cropland base. Manassa would later gain fame when Jack Dempsey, the son of quiet Mormon farmers, became a world boxing champion known as "The Manassa Mauler."

In the 1890s, a representative of the Holland American Land and Immigration Company toured the San Luis Valley and bought 15,000 acres near Alamosa. The Empire Land and Canal Company also actively wooed settlers to the area. Most of the Dutch settlers quit after a few months, worn down by the isolation, hard work, and financial chicanery of the real estate companies. A few families took up homesteading in other parts of Colorado, but most returned to Holland broke and disillusioned.

The valley's irrigation companies began to greatly improve their canal systems during the 1910s, which provided splendid opportunities for agricultural expansion. During the 1920s, sugar beets were widely cultivated, and the population of Saguache, Rio Grande, and Alamosa counties was rising sharply. Heavily Hispanic Costilla County did not share in this prosperity and has continued to lose population since the 1930s. Those who remain work as seasonal laborers on Anglo vegetable farms where they work beside migrant pickers from rural New Mexico towns, Mexico, and the Navajo and Jicarilla Apache reservations.

Today, the San Luis Valley produces 85 percent of Colorado's $60 million annual potato crop. Significant amounts of barley, oats, melons, and vegetables are also raised locally. The valley's population is 40 to 50 percent Hispanic, with Spanish as the primary language in a third of the valley. Colorado as a whole is only 12 percent Hispanic.

The San Luis Valley is a persistent pocket of hard times. Twice as many residents as the statewide average live below the poverty line. Most of these people are named Medina, Pacheco, Rivera, Vigil, Gallejos, Martinez, or Valdez. One part-time resident who lived decidedly above the norm was named Forbes.

The late Malcolm Forbes owned and operated *Forbes* magazine, which had been started by his father in 1917. This glossy, biweekly publication focuses on high finance, big business, and luxurious travel. Malcolm Forbes was the model for this lifestyle. He served in World War II and was awarded the Bronze Star and the Purple Heart. Until his death in 1990, he managed diverse investments, designed and flew castle-shaped hot air balloons, owned seventy Harley Davidson motorcycles and twelve Faberge eggs (the Russians have only ten), and traveled the world in a jet named *The Capitalist Tool*. Like so many others before him, he also dealt in real estate in the San Luis Valley.

The Sangre de Cristo land grant, which covers a large part of the San Luis Valley, has a convoluted, byzantine history. In 1844, the Mexican government granted a vast but vaguely described area to Charles Bent, the governor of the New Mexico colony, and to Stephen Luis Lee. During the 1847 Taos Revolt, Pueblo Indians killed and scalped Bent and hacked Lee into small pieces. Charles Beaubien inherited Bent's share and paid $100 to Lee's heirs to consolidate his title.

In 1851, San Luis was settled to establish an American presence in the valley following the 1848 Treaty of Guadalupe Hidalgo which ended the Mexican War. By 1853, subdivision of the Sangre de Cristo land grant began. One-sixth interests ended up in the hands of French fur trapper and trader Ceran St. Vrain and others. By the time the U.S. Surveyor General's Office in Santa Fe confirmed the legitimacy of the grant, the title was scattered into many hands. Yet the property boundaries remained unsurveyed.

In 1861, Colorado became a separate territory. Thereafter, one-quarter of the grant was part of New Mexico. Charles Beaubien, the principal owner, publicly announced his intention to sell. Future Colorado Governor William Gilpin knew the region from his experiences on an expedition with Fremont 20 years earlier. In 1862, Gilpin bought a three-month option and went to California and New York to raise the needed purchase money. Despite numerous financial setbacks and a fractured land title, Gilpin eventually owned all the grant's 1 million acres.

William Gilpin was a born optimist. In the mid-1860s, he formed the Colorado Freehold Land Association to promote the area. It wasn't easy. The valley had the reputation of being vastly underdeveloped, with no railroad and too many angry Indians. In frustration, Gilpin traveled to London looking for investors. William Blackmore, a London financier, and Charles Lombard, an attorney for the Union Pacific Railroad, were brought to Colorado to handle land sales. A $15 million price was advertised for the whole grant, which two decades earlier had been worth perhaps a few thousand dollars.

An Englishman like Blackmore had trouble comprehending a piece of property the size of Devon and decided to split the grant into two pieces. The subdivision occurred roughly along the Culebra River. The south half became the Costilla Estate, and the north half became the Trinchera Estate. His company also changed its name to the Colorado Freehold Land and Emigration Company to better promote the grant as an international real estate attraction. Company pamphlets described nonexistent surveys of the upland into parcels 2 to 5 square miles in size and the bottomland into 40- to 160-acre agricultural tracts.

In 1870, a Dutch Company, the U.S. Freehold and Emigration Company, bought the 500,000-acre Costilla Estate. The Trinchera had poorer water resources and proved impossible to sell. Settlers could acquire 160 acres of better land nearby for free under the Homestead Act. The Denver and Rio Grande Railroad reached Alamosa in 1874, lessening the valley's remoteness but not Blackmore's struggle. By the 1880s, Mormon farmers were thriving on the irrigated benches across the Rio Grande from the still empty Trinchera Estate. William Blackmore grew increasingly frustrated and morose. He eventually took his own life.

In 1907, a group of investors from Colorado Springs acquired the 500,000-acre Trinchera Estate. They dreamed of promoting a land boom to capitalize on America's growing nostalgia about going back to nature. Reservoirs, irrigation canals, and other developments were planned, but most of them were never built. People simply weren't interested in settling in a distant, heavily Hispanic Colorado basin.

The investors soon subdivided out a 263,246-acre parcel and called it the Trinchera Ranch. It contained dry sagebrush grasslands on its lower slopes, pinyon-juniper woodlands on the foothills, and commercial forests at higher elevations. The land teemed with elk, deer, bear, mountain lion, and other wildlife. In 1913, a wealthy polo player from Denver named David Bryant Turner bought the ranch only to lose it in the 1929 stock market crash.

When Ruth and Albert Simms of Albuquerque bought the property in 1938, it was the largest remaining mountain ranch in Colorado. Land title problems forced the Simmses to initiate a "quiet title" legal action to clear their ownership. The case involved over 400 people and corporations. When their land title was finally firmed up, the couple settled in to harvest the forests and raise horses, herefords, and buffalo. Following the death of the Simmses in the mid-60s, national magazine ads offered the ranch for sale. The heirs, facing estate tax burdens, sold the vast entirety of the Trinchera Ranch to Malcolm Forbes for $3 million.

At first, Forbes tried raising beef by the Japanese kobe system, which requires a special diet and daily massages for the cattle. This outlandish scheme to produce melt-in-your-mouth tenderloin proved to be a reliable money loser. Forbes then approached the Colorado Division of Wildlife (DOW) with the idea of creating the largest private game preserve in the United States. The notion was for the DOW to study the interactions of range livestock with elk and other big game. Private hunting permits would be sold by the ranch based on the game management recommendations of state biologists. The project went nowhere. Under Colorado law, wildlife belongs to all the people of the state and cannot be fenced as Forbes had intended. In 1978, *Forbes* magazine's Ruth Armstrong wrote:

> *Only then did Mr. Forbes turn to real estate subdivision. So far, the company, Sangre de Cristo Ranches, Inc., has sold over 7,000 parcels of land ranging upwards from 5 acres each, or a total of 60,000 acres. (Armstrong, 82)*

In the late '70s, a 13,000-acre area of one-acre mountain tracts was offered for sale under the name Forbes Park. In 1989, a corporate ad appeared in numerous national magazines offering 17,000 acres divided into "Forty acres of America the Beautiful at the new Forbes Wagon Creek Ranch." The text informs us this is:

> *One of the last remaining ranches in America. In season you can hunt deer, elk, bear, wild turkey, grouse, and other abundant wildlife. Many of the 40-acre ranches command an unobstructed view of Mount Blanco (14,345'), the fifth highest peak in Colorado.*

To the Navajo, Mount Blanco is a sacred mountain which delineates the northeast arc of their world.

Prices for lots in the Forbes Wagon Creek Ranch start at $25,000, in the Sangre de Cristo Unit at $4,500, and in Forbes Park at about $6,000—all with as little as a 1 percent down payment. The often substantial balance is financed at a 9 percent annual rate for varying periods of years. At the Forbes real estate office on the Trinchera and out on the land, you learn why the parcels are sold so cheaply. The substantial depths to groundwater can raise the cost of drilling a well above the price of the tract itself. In many areas, the infrastructure is marginal—mostly primitive roads cutting through the forests and meadows. Utility services are not available to large portions of the project.

Few houses have been built on the Trinchera Ranch. Many sales are made to out-of-staters who buy on impulse and never get around to doing anything with their land. The Forbes real estate literature bears a standard, legally mandated disclaimer in bold print which warns the reader to:

> *obtain the property report required by Federal law and read it before signing anything. No Federal agency has judged the merits or value, if any, of this property.*

The Costilla County government has created no zoning districts in its jurisdiction and has had little or no impact on the subdivision of tens of thousands of acres. Locally, the sale of real estate is believed to be a form of economic development. Think again. The "town" of Fort Garland was to be the main service center for the Trinchera Ranch settlers. Today, it is a re-fueling whistle-stop on Route 160 where dogs nap next to the railroad tracks running beside a single small grocery store. Driving through much of the subdivided portions of Forbes's Trinchera Ranch gives an eerie feeling of traveling through a ghost town. The San Luis Valley is once again shrugging aside another misguided attempt to claim it. Only the residents of the Alamosa area, including 3,500 or so Mormons, and the few thousand Hispanos in the San Luis area south of the ranch are sticking it out. It is baffling why a bright, wealthy man such as Malcolm Forbes chose to ignore history and pursue the dream of a land boom. Perhaps it is because he saw the ranch as simply another toy. The ranch cost $3 million. He spent $2 million in Morocco on his seventieth and last birthday party.

LA GARITA RANCH

In 1982, the Colorado Field Office of The Nature Conservancy (TNC) contracted the Montana-based firm of Bruce Bugbee & Associates to prepare an ecological baseline study for the 1,400-acre La Garita Ranch on the upper Rio Grande near the old silver mining town of Creede. The Phipps family of Denver, owners of the property since the turn of the century and one-time owners of the Denver Broncos, wished to protect the land's ecological and scenic values by donating a conservation easement to TNC. Michael Stewartt, who operates a conservation flying service called Project

Lighthawk out of Santa Fe, flew us in as consultants to do a field survey of the ranch, map and analyze its vegetation and landforms, make wildlife observations, and provide recommendations for the design of an easement.

The ranch had been bought by Senator Lawrence Phipps in 1908 and, after a short-lived attempt at dairying, was used continuously for recreation and limited cattle grazing. The La Garita Fishing Club was soon formed. Four ponds were built along Bellows Creek in the center of the ranch, and cabins were put up for flyfishing guests. Allan Phipps, son of the senator and one of Colorado's wealthiest men, spoke knowledgeably about the role of the ranch as the winter range of a large elk herd. He produced a photo showing 354 elk in one pasture during a recent winter, proudly counting out each one. Best take him at his word. He is a man used to dealing with numbers.

Allan Phipps is also a gifted rockhound with an extraordinary collection of ore samples, minerals, and volcanic ejecta. Growing up on the La Garita Ranch no doubt had its influence. The property lies within southwest Colorado's San Juan volcanic field. The present physiography of the horseshoe-shaped upper Rio Grande Valley is the product of an astonishing sequence of geologic upheavals over the last 30 million years.

For millions of years of Tertiary time, molten magma in underground chambers triggered episodes of spectacular volcanic activity across the region. Ash-flow and lava eruptions spewed successively from a suite of volcanos. Lavas, welded tuffs, and diverse pyroclastic deposits piled up chaotically on the surface. As the supply of molten rock decreased, volcanos collapsed into their underlying magma chambers and explosively disintegrated.

In *Basin and Range*, John McPhee describes what the scene may have been like:

> *Up through hundreds of fissures, dikes, chimneys, vents, and fractures came violently expanding, exploding, mixtures of steam and rhyolite glass. Enormous incandescent clouds, heavier than air and hot enough to weld, moved across the landscape like scalding dust storms, covered streambeds and valleys, and obliterated entire drainages like plaster filling a mold. (59)*

In late Tertiary time, the Creede Caldera collapsed and detonated. A moat formed in its fault-ringed margin where interbedded deposits of travertine from hot springs and sediments from lakes and streams accumulated.

The Rio Grande, Bellows Creek, and other streams have excavated their channels through this softrock smurge. As a result, the Creede Caldera is nearly encircled by watercourses rendering the landform visible from space. Pleistocene glaciers provided the finishing touches on the landscape by creating cirques, horns, cols, and aretes in the rugged San Juan and La Garita mountains.

The region's biology is as epic as its geology. The upper Rio Grande Valley and downstream San Luis Valley possess rich and nationally important wildlife habitats. Elk, bighorn sheep, bobcat, fox, and mountain lion roam the landscape. Raptors such as bald and golden eagle, prairie falcon, peregrine falcon, goshawk, red-tailed hawk, kestrel, great-horned owl, and saw-whet owl fill the sky. Census counts in the San Luis Valley indicate that during winter there are 16 bald eagles and 11 golden eagles per 100 square miles or totals, respectively, of 300 and 200 birds.

The La Garita Ranch region is a key eagle wintering habitat where birds feed on rainbow, brown, and Rio Grande cutthroat trout. The upper Rio Grande has been rated a "Class I—Major Fishing Stream." Wintering and migratory sandhill cranes and black-crowned night herons are common. During some years, 14,000 sandhills fly into the valley. Whooping cranes are even honored by an annual festival in Monte Vista. Waterfowl abound in the San Luis Valley wetlands such as the Alamosa and Monte Vista National Wildlife refuges. Many species move up the river onto the ranch. Pintail, mallard, gadwall, cinnamon teal, green-winged teal, redhead, merganser, and Canada geese are present in huge numbers in this portion of the Central Flyway. The Phipps family resides in a truly remarkable place.

Sometimes the prime rationale for a conservation easement is changed because of what is discovered in the field. The La Garita project was designed to chronicle the Rio Grande fishery and the presence of elk winter range, and raptors, and to investigate the possibility of using the property as a site for the reintroduction of river otter. It turned out the ranch also contained excellent Arizona fescue/mountain muhly grasslands. This assemblage has been classified as a "Plant Community of Special Concern in Colorado" by the State Department of Natural Resources. Today, The Nature Conservancy highlights these unusual grasslands in its literature on statewide projects.

The Phipps family conservation easement was quite restrictive. The existing cabins can be repaired, the ponds maintained, and livestock grazed, but the 1,400-acre ranch cannot be subdivided for housing tracts. The federal income tax deductions made available to achieve this was money well spent.

Southern Colorado: The Outback

In recent years, the owners have spent several hundred thousand dollars on stream restoration work.

It should be noted the family has subdivided 92 adjacent acres into 28 parcels and is marketing them under the name Wagon Wheel Gap Estates. Allan Phipps's son, Tony, is in charge of the project. However, the protection of La Garita was far more than a cold-blooded tax shelter. The income tax deductions and potential estate tax savings, while substantial, were far less than the family could have gained from an unrestricted sale of the land.

The Phipps's wealth is sizeable and long-standing. Allan has, as locals put it, "pissed with the big dogs" for decades. Money doesn't explain the conservation easement gift. Here was someone who already owned an NFL franchise. At the end of the field work, the truth came out. Allan Phipps talked about how much he cared for ex–Denver Broncos linebacker Randy Gradeshar because of his loyalty. He then reminisced about his distant childhood spent fishing on the ranch with his dad. The land was a repository of emotional family memories. At last, this elderly entrepreneur looked out over the Bellows Creek meadows awash in blue iris and quietly revealed that this conservation easement was the most important deal he had ever done. To him it was an act of loyalty. Despite rising land development pressure, no other ranchers in the upper Rio Grande watershed have made a similar gesture of stewardship.

ALAMOSA TO PIKES PEAK

A light plane offers a superb vantage point for exploring south central Colorado. Once, while on a consulting job, as we were approaching the Sangre de Cristo Mountains, all hell began breaking loose on the crestline from a battery of thunderheads. It was an easy call. The pilot chose to land in Alamosa to wait things out.

Leaning against an airport shed and drinking a root beer, a local mechanic recalled the story of an unfortunate horse named Snippy, who was killed in 1967 about 20 miles east of town. The poor gelding was found with his throat cut, the flesh and hide stripped from his skull, and his brains and spinal fluid gone. What caused this mutilation was a source of heroic levels of wildass speculation. Local opinion finally resolved that crazed aliens from UFO MASH units were to blame. A vet later autopsied the animal and concluded it had been near death from an infected flank when someone mercifully slit its throat. Predators and baking sun did the rest. An enterprising local man finally claimed the skeleton and trussed the bones together with metal rods. A reassembled Snippy was set out in a place of decorative honor beside the parking lot of a pottery shop west of Alamosa.

Back aboard the Cessna, the pilot flew low over the Great Sand Dunes National Monument. The wind buffeted and bounced the plane like a pollen grain. This sort of ferocious westerly wind fetched up a sand sea at the end of the Ice Age when glacial sediments lay bare in the San Luis Valley. However, local lore leans toward a far different genesis for these dunes, which can reach over 700 feet in height.

There is a tale of a man named Don Louis Gonzales, who entered the San Luis Valley in 1816 with a party of fellow Mexicans and a large flock of sheep. The group built cabins and settled in. Then a massive windstorm supposedly swept up from the southwest and buried the men, their cabins, and 3,500 sheep in 50 feet of sand. As the story goes, the dunes have grown to their present size above these grisly graves.

From the air, circles from the center-pivot irrigation systems repeated out to the horizon. East of the Sangres, a rectangular land survey grid dominated. Pueblo and the wheat fields and grazing lands of the eastern Colorado plains lay off in the haze. This community of 135,000 anchors the southern end of the Front Range sprawl.

Fort Pueblo was built in the 1840s and served most notably as a transshipment point for Taos Lightning—the strong beverage of choice among miners all over the Central Rockies. Flying northeast, a smudge from the smokestacks of the city's steel mills trailed westward up the Arkansas River Valley. Water for these mills, the city of Pueblo, and the agricultural landscape comes from the Frying Pan–Arkansas Project, which begins west of the Divide at Aspen.

Pueblo is home to the Southern Colorado Heritage Conservancy. This land trust was incorporated in 1988 and received its non-profit status in 1989. So far, it has focused on historic preservation projects, but biological conservation is also part of its charter. A trust member explained, "We are best at being an enabling land trust—we like to bring groups together." With no professional staff, all legwork and negotiations are done by board members. A local woman points out what a drawback this is "when things occur like my hip replacement surgery and the death of the board Chairman's brother in a ranching accident."

Despite the setbacks, the trust has three main projects underway. It is working to restore 1860s-vintage Territorial-style houses in a hamlet far down the Arkansas near Lamar. At Agillar, south of Walsenberg down I-25, the group is attempting to move the historic Foster Stage station to a site next to the town's museum. Its most ambitious project involves restoration of a large barn on the former ranch of cattle baron Charles Goodnight, near Pueblo. He was famous for the Texas cattle drives along the Goodnight-Loving Trail.

Five groups, including the National Trust for Historic Preservation and the Colorado Department of Parks and Recreation, are involved in trying to acquire the Goodnight Barn from a gravel company. The Southern Colorado Heritage Conservancy also is interested in protecting several remote sites with pictographs and petroglyphs, a stage station near La Junta, and a large exposure of dinosaur tracks along the Purgatoire River currently used by the Army for maneuvers. The trust hopes to do more near Pueblo, but without a staff, it is ill-equipped to contend with the growth and development pressures around the margins of the city.

Nearby, the town of Cripple Creek squats on the western flank of Pikes Peak. This little burg was the site of one of the richest gold strikes in Colorado history. The complex geologic structures of the mountain were at first thought unpromising. The discovery of the Cresson Vug in 1891 proved otherwise. Some of the ore taken from this vein system assayed at $100,000 a ton.

One of those who profited from this windfall was Winfield Scott Stratton. Stratton made a $10 million fortune, but had visions of even grander auric splendor. He noticed that the main shafts of his Independence and Washington mines were all angled toward a vortex deep within the Earth. Stratton believed these rich lodes were only small offshoots from what had been a cauldron of nearly pure molten gold. He called it "The Bowl of Gold." Test shafts proved inconclusive, and Stratton died in 1902 with his dream unfulfilled.

During the settlement era and mining booms, Pikes Peak served as a beacon welcoming plains-weary newcomers into the safe harbor of the Rocky Mountains. Thousands vowed to make it to Pikes Peak or bust trying. The mountain is named for adventurer Zebulon Pike, a native of New Jersey, who explored the Colorado area in 1806 on behalf of Thomas Jefferson at roughly the same time Lewis and Clark ventured up the Missouri River. The mountain bowled Pike over, and he initially put its height at 18,582 feet. Its actual elevation is 14,110 feet. Pikes Peak visually dominates the terrain west of Colorado Springs and draws skyward the eyes of cadets at the U.S. Air Force Academy. From the air, extensive subdivision road systems can be seen lacing the foothill zone.

Colorado Springs began as the Fountain Colony. That settlement, named for Fountain Creek, was to be far more than a workaday agricultural town. It was supposed to become a "Little London."

William J. Palmer, founder of the Denver and Rio Grande Railroad, was a man with a fascinating dream. In 1872, Palmer formed the Colorado Springs Company as a real estate subsidiary of the D & RG, bought 10,000 acres, platted a townsite, and sold memberships. By 1882, the town had 4,500 residents, many of them newly arrived from London. Palmer believed in wide boulevards, parks, and open space. Colorado Springs was designed to be an island of refinement in the raw West. The city served as the first capital of the Colorado Territory. Manitou Hot Springs, the vertical sandstone hogbacks of the Garden of the Gods, the looming presence of Pikes Peak, and the high, dry air continued to attract people to farm, run businesses, or live off family money in Colorado Springs.

In his book *Newport in the Rockies*, Marshall Sprague writes: "Colorado Springs was created in a high fragrant wilderness of spectacular beauty for the sole purpose of making men and women comfortable" (177). During the 1970s, Colorado Springs would be the ninth fastest growing city in America. Today, most of its 230,000 residents earn decent incomes from high tech firms, the Air Force Academy, the agricultural industry, the University of Colorado campus, retailing, and tourism. Local people unabashedly describe the climate as perfect.

The Palmer Foundation is based in Colorado Springs. This land trust, named after railroad magnate William J. Palmer, has been in existence since 1977. A board member describes it as "basically a Pikes Peak land trust operating in both El Paso and Teller counties." The trust owns 20 acres and holds conservation easements totaling 1,980 acres of protected land. However, its operations are expanding. In early 1989, the organization hired its first paid employee. Additional conservation projects are currently nearing completion. These range from ranchland around the edge of Colorado Springs to small tracts of open space and parkland near the Garden of the Gods.

The Nature Conservancy recently gave the city 5 acres adjacent to the Garden of the Gods for open space and recreational purposes and has secured a protective 99-year lease on 1,000 acres in nearby Aiken Canyon. Airborne you can see the size of the challenge which remains. Roads and houses extend in every direction from the city, but the foothills of Pikes Peak appear to be most threatened.

Flying over this area on one occasion, our pilot coaxed the Cessna into a series of smooth, flawless victory rolls. Passing the summit of Pikes Peak upside down, he howled like a monkey from the sheer joy of it. That mountain has a way of working on people. In 1876, John O'Keefe, a sergeant in the U.S. Signal Corps, was ordered to establish a weather station at the top of Pikes Peak. After spending most of the next three years in oxygen debt, he reported fighting off malevolent packs of screaming lunker mountain rats capable of stripping a side of beef like a school of piranha, feigned the loss of a nonexistent infant daughter consumed by these Paleocene-scale rodents, and telegraphed descriptions of fiery volcanic eruptions and a lava lake pouring molten rock buckety-buckety down the mountain toward the unsuspecting citizens of Colorado Springs. Locals caught on to his genius for crafting an elegant lie and held a rowdy banquet in his honor when he left the area in 1881. Today, Pikes Peak still attracts extremes of human behavior, with an automobile race to its summit (usually won by one of the Unser clan) and a marathon which requires contestants to run up *and* down the mountain. If you survive, you get a T-shirt.

PAGOSA SPRINGS AND DURANGO

Pagosa Springs is the seat of Archuleta County. Juan Archuleta was a powerful figure in the Santa Fe Office of the Court of Private Land Claims, charged in the 1800s with settling land grant disputes in New Mexico and Colorado. The natural hot springs initially attracted Utes and now draw in hordes of tourists. Pagosa Springs is today a lower case resort town far enough up in the mountains that the San Juan River is a clear trout stream.

Subdivision of valley ranchlands is widespread. "Mountain real estate" developments are coming into favor, such as the one marketed by Jasper Properties Inc. just over the Conejos County line. Wolf Creek Pass in the San Juans is the site of ongoing battles over ski development. It was here, in 1979, that the last verified grizzly bear was killed in Colorado. The endangered peregrine falcon is reported to reside in this same contested area.

In 1981, the Trust for Public Land (TPL) delivered its first land trust baby in Colorado. This group was called the Upper San Juan Land Protection Association. TPL assisted the group in procurement of its 501(c)(3) nonprofit status from the Treasury Department and temporarily held a conservation easement which was later transferred to the local trust.

Local support for the group was slow in coming. In 1985, the trust changed its name to the Southwest Land Alliance and broadened its sphere of activity to include much of southwestern Colorado. The Alliance now holds conservation easements on three ranch properties totaling 2,100 acres and on a 75-acre inholding in the San Juan National Forest. Two conservation easements are pending on ranchlands.

Despite the growing list of successful projects, the trust is in danger of folding. The director's annual operating budget is only $3,000. Despite endowment, monitoring, and education bank accounts which total many times that figure, the current board of the Southwest Land Alliance seems unwilling to fund the executive director position.

Durango, the seat of adjacent La Plata County, was created when six homesteaders were bought out by the Durango Land and Coal Company in 1880. The company then platted a townsite over the ranchland. A year later, the Denver and Rio Grande Western bridged the Animas River and entered town. Today, Durango has a population of 13,000 and is the home of Ft. Lewis College. The community, best known as the headquarters of the Durango and Silverton narrow gauge railroad, is the tourist hub of the south slope of the San Juan Mountains.

Refurbished locomotives pull passenger cars full of tourists up the Animas River Canyon 45 miles from Durango at 6,520 feet to Silverton at 9,300 feet. The eight-hour round trip costs $32 per adult. Durango sits literally on the border between the Colorado Plateau and the Rocky Mountains. It is flanked on three sides by sandstone mesas with sparse stands of Gambel oak and unvegetated grey slopes of Mancos shale. The high, glaciated San Juans begin just to the north.

Downtown Durango is a prototypical faux Western shopping district for tourists. The front of the Ralph Lauren factory store has a 15-foot-tall mounted polo player painted on it. Nearby is the Benetton Factory Store, Rosewater's Deli, the Ice Cream Factory, numerous bars with neon cowboy art, the Main Mall, and rows of brick storefront boutiques, which these days pass for historic preservation projects. A bit west of the downtown mall is the Kiva Theatre. Durango is in Southern Ute country, and Utes were enemies of the Kiva-building Pueblo tribes. The Dos Juans Restaurant adds to the confused cultural melange. Painted on the side of the restaurant is a scene where two Mexicans are taking a siesta beneath sombreros as they lean, you would guess painfully, against a large saguaro cactus. Aside from the Dos Juans parking lot, the nearest saguaros are a few frost-damaged stragglers on lower slopes of Arizona's Mogollon Rim, some 250 miles southwest of Durango.

In 1989, *Outside* magazine named Durango one of the premier places to live for those seeking a four-season recreation lifestyle. It's the kind of town where you simply cannot own too much Patagonia clothing. Evidence of a rising demand to live in La Plata County can be seen by driving north on Route 550 up the Animas River Valley toward the Purgatoire ski area. The reddish sandstone cuestas above the cottonwood bottomland support ponderosa pine and Douglas fir forests. The meadows along the river are full of horses and Angus cattle. A few miles from town is a wooden sign for "The Ranch," a large subdivision offering "Waterfront Homes." It's tastefully done with underground utility wires and attractive landscaping. However, tract housing is also beginning to fill the valley.

Up the road a sign placed in a pasture blares "For Sale: Vacant Raw Land. Burns National Bank, Durango." Nearby hay meadows and aspen groves are advertised as "R. T. Scott's Fishing Ranch." Eight "improved" 3-acre homesites are offered at $52,500 each.

In the Animas River Gorge, the sedimentary rocks of the Colorado Plateau form southward-dipping ramps to the Precambrian core of the San Juan Mountains. Near Electra Reservoir, the peaks themselves become visible. At a wide place in the road, the words "Scenic View" have been spray painted on the asphalt with an arrow for emphasis. In the needlegrass meadow ringed by blue spruce and aspen is a metal sign for "God's Country Realty." The Christian fish symbol is used as a dash between the prefix and the last four digits of the telephone number.

The Nature Conservancy has been very active near Durango. It acquired 7,000 acres of elk and deer habitat which later became the Division of Wildlife's Bodo Wildlife Area and also facilitated the addition of 295 acres to the BLM's Perins Peak Wildlife Area, where peregrine falcons have been reintroduced.

A local planner expressed concern over the inability of the county's subdivision regulations to support these preserves:

> *In Colorado, parcels 35 acres or larger are exempt from our review except for septic approval and the issuance of a building permit. Around here though, most subdivision is happening in what we call Category*

*One minor plats where tracts are three acres and above. These plats
don't even go before the Planning Board and are approved by the planning staff, almost as a matter of course.*

There is no countywide zoning, and the local mood largely favors noninter-
ference. However, a new county comprehensive plan is being prepared, and
the emphasis will be on "performance zoning" and negotiated development.
In 1993, due to a 10 percent annual growth rate, Durango declared a six-
month moratorium on new building permits. Whether this will last is a hotly
debated political issue.

The La Plata Open Space Conservancy functions as a private land trust
which started as a part of the planning program. Its brochure outlines what
conservation easements are and explains associated tax benefits which may
be available to landowners who donate easements to the trust. The trust has
completed an inventory of state- and county-owned lands which may be
available to trade for important parcels of open space. The literature points
out that "Today, many demands are being placed on the undeveloped areas
of La Plata County. It is up to the landowners to protect and preserve these
important lands."

The La Plata Open Space Conservancy is an example of a former public
agency which was forced to become a private land trust in order to survive.
As long as the group was an arm of government, local landowners were very
reluctant to talk about land-use options. Now, local confidence is growing in
the ability and honesty of this private conservancy. The trust is striving to
protect scenic corridors along Routes 550 and 160 near Durango and ranchlands in the Animas River Valley. It owns 530 acres and holds conservation
easements on another 540 acres.

The La Plata County Commissioners also played a role in this shift from
public to private sector action. In past years, they raided the Open Space
Fund to pay for improvements to the fairgrounds. Money which could have
been used to protect ranches from suburbanization was spent on staging
rodeos for the tourists. The cowboy myth gallops on.

10

THE WEST SLOPE: RIVERS, OIL SHALE, AND SUNSET LIVING

GRAND JUNCTION

In the 1870s, a massive Ute reservation extended roughly from the Colorado River north to the White River Valley. Covetous whites soon began crowding the Indians out. In 1879, enraged over increasing trespass and decreasing supplies, the Utes made a series of attacks. White River Indian Agent Nathan Meeker, a former agricultural editor for Horace Greeley's *New York Tribune*, 8 employees of the Indian agency, and 36 soldiers were killed. Public outrage over the so-called Meeker Massacre resulted in removal of all Colorado Utes from the state except for those in the Southern Ute Reservation near Cortez and Durango. The Utes were driven from their game-rich hunting grounds in the Piceance Basin and relocated to the high, desolate Uinta Basin in northern Utah.

As a result, a vast expanse of arable lands in the Gunnison and Colorado (then Grand) river valleys opened to white settlers. The Town Company acquired land and began selling tracts where the two great rivers came together at present-day Grand Junction. By 1882, numerous irrigation ditches had made commercial fruit growing profitable. Extensive acreage of orchard tracts was platted and sold. The Denver and Rio Grande Railroad soon reached Grand Junction, which for a time was called West Denver.

Livestock, alfalfa, and vegetables thrived next to the orchards in the warm, sunny valley bottoms. By 1899, enough sugar beets were grown in the Grand Junction area to support Colorado's first sugar mill.

During the late 1940s, a uranium boom in the Colorado Plateau sup-

^^^^^

plied several uranium mills in the growing community with "yellowcake" ore. For years, dust from the radioactive slag piles blew across the city. Some of the slag was used as aggregate in the foundations of post-World War II homes. It is still common practice when renting or buying a home in Grand Junction to have the property tested with a Geiger counter. But the radio-active image, at least, is almost gone. The Uranium Motel has been boarded up, and its neon atom sign no longer glows in the dark.

Today Grand Junction is a community of 30,000, supported by retail trade, railroads, agriculture, and (during upturns in the fossil fuel cycle) oil and gas development. In the 1970s and early 1980s, the city and surrounding Mesa County were engulfed in an energy boom. Commuter flights from Denver to Grand Junction were full of executives working for Texas-based energy companies. Lately, however, the overhead bins of those same flights hold few Stetsons.

Mesa County remains rich in orchard culture. This legacy unfolds in towns such as Fruita, Fruitvale, Rhone, and Appleton and in adjacent Delta County within Orchard City. Peaches, apples, pears, and grapes continue to be grown commercially. There is a winery in Grand Junction that bottles a fairly good Chardonnay to serve with fresh rainbow trout caught from local creeks. The Colorado, Uncompahgre, and Gunnison river valleys in Mesa and Delta counties are still the most important fruit-growing districts in the state. Pinto beans—spotted, protein-rich, and perfect for refrying—are a major vegetable crop. But the South Platte Valley and the extreme southwest corner of Colorado, near the narcoleptic hamlet of Dove Creek, grow more.

Grand Junction is the only urban center between Denver and Salt Lake City. Land development pressure is steady, driven by natural population gain in the metro area and second home subdivision projects near the forested recreation areas of Grand Mesa. However, during oil and gas drilling fren-zies, intense demands arise for rental housing and trailer park pads.

In 1980, the Mesa County Land Conservancy was formed to protect local orchards, ranchlands, and wildlife habitats from subdivision. Today, the group holds conservation easements on 400 acres of orchard lands. How-ever, other parcels are being evaluated for protection via the Farmers Home Administration (FmHA) Section 118 Farm Debt Restructuring and Conser-vation Set-Aside Program.

Under this voluntary system, FmHA forgives delinquent farm debt on wetlands and highly erodible lands in exchange for placing conservation ease-ment restrictions on the property. The easements prohibit plowing, draining, or splitting the land for non-agricultural purposes. For a farmer to qualify, the loan must have closed prior to December 23, 1985 (the date of the Food Security Act), and the land must have been row-cropped or otherwise cultivated. FmHA's plan is to forgive loans which are unlikely to be repaid in exchange for conservation of delicate lands. The fine points concerning FmHA's easement negotiations, monitoring, and enforcement are still being worked out nationwide. With 118,000 delinquent FmHA borrowers eligible, this is a program worth watching.

At present, the Mesa County Land Conservancy is the only land trust in the country using this technique. A trust member is realistic about the group's chances. "We need to protect at least 2,000 acres of orchards to keep a viable fruit industry in the valleys," he said, "so we've got our work laid out before us." The Mesa County land-use planning office recently attempted to implement a set of regulations designed to limit conversion of agricultural land to other uses. The locals reacted like they'd been gutshot. Another trust member believes that "the trust was wise to steer clear of being associated with that mess—we're not out to regulate anybody."

The Mesa County Land Trust also holds the largest conservation easement in Colorado on the 4,000 acres of the Big Creek Ranch on Plateau Creek, a tributary of the Colorado River. The ranch consists of three parcels near Colbran in the Plateau Creek Valley east of Grand Junction and north of the Grand Mesa. The mesa is a basalt-capped highland with average elevations above 10,000 feet. Over three hundred lakes are interspersed with subalpine meadows and spruce forests. Landslide deposits ring the mesa scarp. Intermediate benches are covered with ground moraine deposited by Pleistocene glaciers which descended from an ice cap on the upland. In the valley, the shale, conglomerate, marlstone, sandstone, and limestone of the Tertiary Wasatch Formation are reservoirs for large accumulations of oil and natural gas.

In the early 1860s, the first settlers entered the Plateau Creek Valley. Among these was Heber Young, a nephew of LDS prophet Brigham Young. Young ranched on what is now called Mormon Mesa west of the Big Creek Ranch. Range wars broke out in the 1890s between cattle ranchers and sheep herders. Several people and hundreds of sheep were killed. The cattlemen formed a vigilante group much like the Sheepshooters Association in Richard Brautigan's book, *The Hawkline Monster.* Their bylaws were basically: "It's OK to shoot sheep." Ever sensible, the Mormon sheep raisers moved on. To this day, cattle ranchers here and in many parts of the West call sheep "meadow maggots."

Colorado: The Icon

Spurred by the Powderhorn ski area and an increasing summer recreational demand on the Grand Mesa, land subdivision is on the rise. While the Plateau and Colbran gas fields now contain only about fifty producing wells, the next energy boom will bring on another, stronger pulse of development. Nearly the entire valley has been leased for oil and gas exploration and development.

The sprawling Big Creek Ranch is a cattle operation which runs both steers and cow/calf pairs. Three producing natural gas wells have been drilled in lower elevation areas of the property. The ranch is the only one left in the valley that contains a continuous corridor for wildlife migration from Plateau Creek at 5,000 feet to the crest of the Grand Mesa at over 10,000 feet. Diverse habitats form a mosaic in this broad ecological span: sagebrush grasslands, pinyon/juniper woodlands, meadows, wetlands, Gambel oak thickets, aspen groves, and spruce/fir forests. The pastures and oak communities serve as critical winter range for hundreds of deer and about 200 head of elk.

Mesa County is the second most hunted jurisdiction in Colorado for big game and birds. The ranch's stands of oak are one of the top wild turkey habitats in the district. During hard winters, up to 60 birds flock to the ranch. Black bear and bighorn sheep inhabit its remote corners. Small mammals include beaver, ringtail, muskrat, raccoon, and weasel. Mountain lion, coyote, red fox, and badgers are the ranch's main four-legged predators. Hunters take the two-legged crown. Turkey vulture, goshawk, red-tailed hawk, prairie falcon, and great-horned owl dine on the ranch's plentiful ground squirrels, gophers, rabbits, and hares. Cottonwood and Big creeks descend from the Grand Mesa and flow through ravines across the property to Plateau Creek, which has been rated by the Colorado Division of Wildlife as a Class I fishery resource.

A perpetual conservation easement has been placed on 4,000 acres of the 6,940-acre ranch. A prohibition on land subdivision was the main stipulation. While the existence of natural gas wells is not particularly beneficial to the ecological integrity of the property, they are a short-term inconvenience compared to permanent conversion of the ranch to housing. The maximum lifespan of a gas well is about 20 to 25 years. Suburbs are forever.

THE GUNNISON RIVER

The rolling sagebrush and pinyon terrain northeast of Montrose is remarkable only for its blandness. The serviceberry, Gambel oak, Oregon grape, rose, and lupine which grow there are common throughout much of the

West. But as you walk across the uplands, there is a looming emptiness more sensed than seen. A moment later, the Earth falls away in front of you. The Black Canyon of the Gunnison has no welcoming sandstone vistas. Your first good view of this national monument comes only when you are two strides from skydiving into it.

The chasm carved by the Gunnison River ranges from 1,750 to 2,750 feet deep. It is as vertical a dark rock canyon as the Barranca Del Cobre in Chihuahua, Mexico. In places, a Little Leaguer could throw a rock from the rim and hit the river. Meandering cream-colored pegmatite dikes have intruded throughout the brownish-black gneiss which gives the canyon its name. Hanging valleys of intermittent tributary streams are the only way to enter the canyon in many areas.

The Gunnison River has substantial motive force. The average grade within the national monument is a stout 95 feet per mile, but over the reach from Pulpit Rock to Chasm View, the river cascades 480 feet in 2 miles. All but truly manic kayakers avoid certain portions of the Gunnison. There are too many rocks and few saving pools. For the inexperienced, this is death water.

The vertical walls of the gorge do provide remote life-saving habitat for eagles and endangered peregrine falcons. The Colorado Division of Wildlife (DOW) recently negotiated deed restrictions in the Gunnison Basin which will prevent 800 acres of sage grouse "booming" grounds from being plowed. This is one of 15 such DOW habitat protection projects carried out in recent years as part of the same Farmer's Home Administration program employed by the Mesa County Land Conservancy. The DOW's first such project was the conservation of 2,000 acres of lesser prairie chicken habitat in southeast Colorado.

The Gunnison is the focus of an ongoing protection effort of TNC as part of its "Rivers of Colorado" program. Water rights attorney Bob Wiginton works in TNC's Colorado Field Office in Boulder. Wiginton believes that over the years TNC has been buying riparian lands without paying attention to water rights. He thinks that it's time for TNC to consider water alone as a subject for acquisition as a resource worthy of the protection efforts of a land trust.

In recognition of this oversight, Huey Johnson, a former cabinet head under California Governor Jerry Brown and one of the co-founders of The Trust for Public Land, has formed the Water Heritage Trust. The goal of this California-based organization is to maintain river flows against threats such as excessive municipal, industrial, and irrigation withdrawals, and dam and reservoir construction projects.

In the West, TNC has begun taking these concerns into the field. It has participated in fisheries improvement work on the Colorado River, reserved 300 cubic feet per second (cfs) of flow in the Little Snake River (a tributary of the Yampa), and received a donation of senior water rights for the Aravaipa River as it flows through a wilderness area near Oracle, Arizona. Yet the Gunnison is the setting of their most intriguing success.

Over the years, the Gunnison River has attracted swarms of dam building and water development enthusiasts. In 1902, the Reclamation Service, now the Bureau of Reclamation, was created as a water development agency. A year later, the Uncompahgre Project on the Gunnison River was authorized as one of the Service's five initial undertakings. Engineering studies showed that a tunnel could be built from the Gunnison to carry huge quantities of irrigation water to the dry sagebrush grasslands of the Uncompahgre Valley. The Gunnison Tunnel was begun in 1904 east of Montrose near Cimarron. Five hundred men labored underground at the peak of construction.

On September 22, 1909, President William Howard Taft's massive form lumbered to the front of a ceremonial platform in Montrose to dedicate the "completed" project. As Taft bellowed to the huge crowd, workmen were secretly holding back seepage water in the tunnel so that when the president called for it, some flow would arrive. The crowd roared and bells were rung as the tunnel emitted a belch of water which subsided as soon as the celebrants dispersed. It was not until three years later, when a diversion dam was built across the Gunnison, that water actually flowed through the 5.8-mile tunnel to the fields of the Uncompahgre Valley.

In 1937, the 200-foot-high Taylor Park Dam plugged a tributary of the Gunnison, the Taylor River, which drained the cirque basins of the Collegiate Range. In 1965, the huge Blue Mesa Reservoir was backed up behind another dam on the main stem of the Gunnison, upstream of the national monument. Today, the Gunnison Tunnel and the rest of the Uncompahgre Project irrigate 86,000 acres of productive agricultural land around Montrose and Delta.

The prodigious vertical drop of the Gunnison River has in recent years led civil engineers to fantasize libidinously over three more dam projects. One of these would back up an impoundment into the Black Canyon of the Gunnison National Monument. Hearing the news, TNC and the State Board of Natural Resources (DNR) gasped and began to assemble strategies for stopping the dams and reserving instream flows. They faced an immediate legal boobytrap. Under existing Colorado law, "water rights" means the right to *consume* water, not allow it to remain in a stream channel. Use it or lose it. So far, no clear precedent exists to turn storage rights into instream flow. Perhaps people have meant it that way.

A prominent Denver water rights attorney argues that such a change can't be made. To him "These instream flow schemes are alarming legally and politically. They [TNC] are trying to let Colorado water slide out of the state. Emotionally for most Coloradans, this is the same as putting it in a pipeline to the Los Angeles Basin." The attorney makes his points with considerable arm-waving:

> *if a conditional water right was granted for storage and power generation, once you change that right to instream flow, you lose all your seniority as a claimant—you become the most junior of all filings. . . .* You've got to remember that water is warfare! *And the new marketers—the water speculators—are drooling over Colorado's water. If we let the water flow out-of-state, these guys will pounce on it and ship it to Phoenix, Salt Lake, and L.A.*

Bob Wiginton of TNC disagrees. According to him, you *can* change a conditional water right without losing your seniority, if it is done correctly.

TNC chose to proceed with the protection of a 29-mile stretch of the Gunnison. First, it negotiated with the Pittsburgh-Midway Coal Mining Company, a subsidiary of Chevron, to waive company rights to build a dam. TNC received a legally binding covenant from the company to that effect. Then TNC worked with the State Board of Conservation to allow Pittsburgh-Midway to donate a 300 cfs conditional water right (worth $7.2 million) to TNC and transfer it from reservoir storage to instream flow. TNC is now attempting to augment this donated flow by negotiating for leased water from upstream impoundments. A member of the Gunnison River Coalition believes the project:

> *has saved the river. However, a week after the Chevron water right gift, the Bureau of Reclamation, with their classic sense of timing, began to solicit comments on a proposal to build a hydroelectric plant on the Gunnison which would divert a significant portion of the river's total flow.*

Colorado: The Icon

Wiginton admits that TNC is making it up as they go. The ground rules for this kind of project haven't even been worked out. Stay tuned.

CRESTED BUTTE

In the upper reaches of the Gunnison River watershed is the former mining town of Crested Butte. While located only 20 air miles southwest of Aspen, high encircling mountain walls make the trip a four-hour drive. Locals like it that way.

However, tourism related to the Crested Butte Ski Area and an associated real estate boom have begun to change this formerly isolated community. In recent years, conservationists have witnessed the towns of Crested Butte and Mt. Crested Butte begin to grow together. The Crested Butte Land Trust was formed to conserve the open space which remains. In 1992, a 1.5 percent real estate transfer tax was passed by the local governments and thus far over $160,000 has been passed on to support the activities of the Crested Butte Land Trust. The trust has used these funds to purchase 40 acres within a former cow camp ranch which has been extensively subdivided. The long-term goal is to buy up additional platted lots to greatly reduce the future density of development. All efforts are fully integrated with the land-use planning programs in the area.

THE PICEANCE BASIN

In Garfield and Rio Blanco counties, life is plain. The rolling sheep ranches of the Piceance Basin produce a steady, if unspectacular, living. The deer hunting is among the best in Colorado, and the winters are generally sunny and tolerable. No wonder the Utes loved it there.

North of Rifle, on Highway 13 heading toward Meeker, is extraordinary country bisected by Piceance Creek. The creek seems underfit to have carved such a deep valley into flat-lying sandstone and a peculiar kind of shale. Pinyon/juniper woodlands cover the slopes, with some ponderosa pine and Douglas fir in shady north-facing pockets.

Several miles downstream a father and son round up some sheep. The man rides a roan horse while carrying a long pole—a sheep poke—in his right hand. The boy follows through the pasture on a purple BMX bike. A border collie named Jeff works the margins of the flock.

Sheep dogs are a serious matter in the basin. In tight times or during a run of hard weather, a good dog can make all the difference. The World Sheep Dog Championships are held in Meeker each year during July. Overflow crowds jam the pastures. A special judge, David McTeir, is flown in from Edinburgh to watch the dogs drive sheep through complex courses of gates, fetch panels, and barrels. First prize is $5,000, and the winners, both handler and dog, get their names engraved on the Meeker Cup. Canadian boys dream of glory on the ice of the Montreal Forum. Rural kids in sheep country fantasize about winning their own kind of Stanley Cup.

The memory of the bloody Meeker Massacre is now Meeker's main year-round tourist attraction. At the White River Museum, details of the battle are dutifully told by cheerful senior volunteers. There isn't a whole lot else to do, even here in the county seat of Rio Blanco County.

A few miles south of Meeker, two Mexican men tend the flocks of a Basque sheep rancher. Years ago, Basques were the seasonal workers, but a number have become landowners, such as the Halandras brothers, for whom these men worked. The Halandras family has created the most unusual hillside letters in the West. Instead of piling rocks or pouring cement, they left a 50-foot-tall "RH" of sagebrush in an otherwise cleared hillside pasture next to Highway 13.

In Rio Blanco and Garfield counties, the land has always reliably supported those who understood it. However, what lies beneath the surface has caused a lot of turmoil. Coal mining occurs north of Meeker near the now busted boomtown of Craig. Oil and gas development proceeds in the Rangely Field. This oil is processed at a refinery west of Grand Junction. Deposits of a natural black asphalt, called gilsonite, exist near Grand Junction in parts of Rio Blanco County and in Utah's Uinta Basin. Gilsonite from this area was the base for the durable black paint of Henry Ford's Model T.

In 1973, efforts to reach deep oil deposits in the Piceance Basin drove the energy industry to bizarre acts. In that year, President Nixon, preoccupied with Watergate, authorized Project Rio Blanco. Atomic Energy Commission and oil industry personnel placed three nuclear devices more than a mile below the Piceance Creek area and, on the morning of May 17, 1973, detonated a 90 kiloton blast which measured 5.3 on the Richter scale. The official explanation for the explosion was that it was to release oil from limestone traps. It did not.

As a result of public outrage over the danger of the blast, Senator Peter Dominick was beaten in the next election by then fresh-faced Democrat Gary Hart, and Republican Governor John Vanderhoof was defeated by the blunt

Democrat Richard Lamm. The Colorado politicians who presided over the Rio Blanco debacle had in effect nuked themselves out of work. All this activity was mere foreplay for what came next.

In the late 1970s, the peculiar kind of shale indigenous to the region was depicted by President Jimmy Carter as America's fuel of the future. Five-hundred-foot-thick deposits of oil shale cover 2,600 square miles of western Colorado. A greasy goo known as kerogen inhabits layers within this other-wise ordinary rock. A ton of oil shale cooked in an underground retort is supposed to produce 25 to 30 gallons of liquefied kerogen. The known oil shale deposits in that part of Colorado have been estimated as equivalent to 2 trillion barrels of oil, more than all the world's known reserves of crude.

The Federal Synthetic Fuels Corporation was formed in the late 1970s as the centerpiece of a government-backed program to fight the Arabs by melt-ing down the Piceance Basin. Exxon, Unocal, and other giants of the energy industry began to invest heavily in oil shale. The Colorado River communi-ties of Parachute and Rifle were chosen as sites for refineries and residential/ commercial developments for the thousands of workers required to get the job done.

A new planned community, Battlement Mesa, was built in 1981 by Bat-tlement Mesa Incorporated, a wholly owned subsidiary of Exxon. The devel-opment arm of the energy giant constructed the "town" in lieu of paying taxes to the government's Synfuels Corporation. By the early 1980s, oil shale leases covered the basin. Just when the boom was poised to happen, eco-nomic reality derailed the whole gambit.

The realization that producing fuel from oil shale was drastically more expensive than from traditional oil coincided with an upswing in the world-wide supply of crude. With Ronald Reagan's election, political momentum favoring alternative energy dissipated, and the project was abandoned. The breaking of the oil shale promise turned many lives upside down.

Today in Parachute, a hand-painted diagram of the region covers a Chamber of Commerce billboard next to the I-70 off-ramp. In the upper right corner, a figure is shown dropping into town suspended by a red, white, and blue parachute bearing the message, "Oil Shale USA: Parachute and Battlement Mesa." Beside the sign is a two-ton block of oil shale donated by Unocal with a sign exclaiming, "Oil Shale—The Rock that Burns!" Un-ocal sporadically operates a test facility north of town. Despite the calamities, the dream hasn't died.

Parachute is a Rube Goldbergesque contraption of new gas stations and shopping areas amid a clutter of run-down buildings. In the neighboring hamlet of DeBeque, the Odd Fellows Hall is shuttered up, and a decrepit boardinghouse with a sagging porch has "4 Sale/Lease Purchase—All bids O.K." sprayed on its side in blue paint.

In Rifle if you ask about oil shale, you tend to get sarcastic laughs or suspicious glares. At a convenience store, a fortyish man says bitterly, "Those bastards promised us real work for decent money. We moved here and got screwed." A guy at the gas station grunts, "We came out better with that crazy artist." This town of 2,000 gained international recognition in 1972 when Bulgarian artist Christo, famous for his California "running fence," hung a 365-foot tall orange nylon curtain across Rifle Gap. The project cost Christo's backers over $700,000. Within a day, mountain winds shredded the curtain and sent a shower of fluorescent orange confetti over the countryside. Despite busted oil shale and artistic dreams, Rifle remains a comfortable small town. But as with a jilted lover, acrimony festers just beneath the surface.

It is only in the planned community of Battlement Mesa that any sort of economic carry-over shows from the oil shale siren song. This new town covers several thousand acres of sagebrush north of the towering landform for which it is named. Even the mesa itself is unstable. Landsliding is common on its upper slopes of buff-colored sandstones and breakable shales.

Downtown Battlement Mesa consists of a single story office complex with a brown metal roof designed to simulate Spanish tile. It houses a grocery store, post office, liquor store, insurance office, the Battlement Mesa Metro District offices, and the headquarters of Battlement Mesa, Inc. The sign on the corporate office reads "Battlement Mesa Information Center and RV Park Check-in."

Inside, the four seasons of life on the mesa are portrayed in large, plastic, backlit display photos. People are shown golfing, riding bicycles, taking jeep tours, and playing tennis. They are all older people. Battlement Mesa, which was to be a town of young oil workers, is now marketed as a retirement colony and recreation playground. The October 31, 1987, issue of *The Rocky Mountain News* contained a full-page article entitled, "Sunset Living: Battlement Mesa Attracting Retirees." An immaculate new community center stands upslope of the shopping area. The town's golf course buzzes with elderly duffers racing electric carts. A player on the eighth tee slices one into the sagebrush. The mesa is winter range for seniors.

Battlement Mesa is like a full-scale land-use planner's mock-up. The community contains areas for modular homes (trailers), condos, and single-family homes. Utilities are underground, and the rest of the infrastructure is equally first-rate. Retirees are urged to select from several "Neighborhoods of Choice." The main street of the weedy mobile home area is appropriately called Tamarisk Trail. The cramped rows of identical tan and brown trailer houses are reminiscent of Daly City, California. Parked out front are the Dodges, Chevys, Fords, and Oldsmobiles of retired couples from cities such as Dayton and Hibbing. Imports are not welcome here. Nearby a field of empty trailer pads lies at the end of the Rainbow Trail. Lines of utility hookup boxes stand next to never-used driveways overgrown with weeds. Several paved cul-de-sacs with red fire hydrants are surrounded only by rabbitbrush and sage blowing in the breeze.

The effort to repackage Battlement Mesa as a real community has largely flopped. A retirement and recreation development was Plan B for Exxon, and it shows. While the project is well-designed, it was conceived as an energy camp and retains an air of empty transience. The corporate logo for Battlement Mesa is enigmatic. Is the sun rising or setting? For now, the place is a polite, ghostly elderhostel. But as you stare out at the surrounding escarpments, you get the feeling that Exxon is simply biding its time until the booms from oil shale retorts once again echo across the Piceance Basin.

THERE
LIES A CITY

Throughout Colorado's history, as the reverberations from one economic boom fade, a louder and grander explosion has always replaced it. Denver and the Front Range corridor, born a network of instant cities, are now an urban amalgam. Aspen and a score of old mining towns have become nationally and internationally renowned ski resorts which empower unrelenting real estate industries. In other parts of the state, land development occurs in pockets around energy resources and recreational amenities. The marketability of Colorado realty to newcomers is enhanced by the state's mystique as an idyllic mountain paradise—as an icon worthy of blind veneration.

The reaction of Colorado's conservationists to the continuing loss of open space and natural ecosystems has been remarkably similar to that of their East and West Coast kin. Twenty-seven land trusts, several national organizations, and cities and counties are at work trying to save the best of what's left of Colorado. This response is, at least structurally, more typical of coastal regions than the remainder of the Rocky Mountain West, where only 16 other land trusts exist.

As of 1992, Colorado's 27 land trusts have conserved 42,873 acres, with over 30,000 of these acres protected by Colorado Open Lands. Recently, members from several local groups formed the Colorado Coalition of Land Trusts, a statewide equivalent of the national Land Trust Alliance. The rapid proliferation of trusts has created a need for improved communication. Most groups are young, function without professional staffs, have meager budgets,

^^^^^

and are held back because of these limitations. Given time and greater use of the statewide information exchange system coordinated by the Colorado Coalition and the Trust for Public Land, the effectiveness of Colorado's trusts is sure to increase. However, it seems clear that for a trust to really succeed, there should be a person hired, even part-time, to attend to business. Consolidation of some of the state's trusts might eventually have to be considered in order to pool resources and to avoid fundraising competition and jurisdictional problems. While Colorado Open Lands is a first-rate and proven organization, there may be room for a West Slope counterpart. This group might be formed from a consortium of existing small trusts which are encountering difficulty gaining momentum. The Nature Conservancy has been very active. By the end of 1992, TNC had completed 61 projects in the state totaling more than 70,000 acres.

In addition, Denver, Fort Collins, Colorado Springs, and several other cities have established extensive parks and open space systems. The open space acquisition programs in Boulder and Jefferson counties alone have conserved over 72,000 acres using the private-market, nonregulatory approach advocated by trusts.

In Colorado, no immovable impediments exist to both land development and land conservation. While it is incontestably true that the profits to be made from subdivision are enormous and seductive, this can also work to the benefit of conservationists. Land-saving is only spoken of where there is land-losing. The immediacy of the threat itself is why many landowners are motivated to conserve.

Nearly everywhere in the Rockies will be "discovered" someday. For decades, ranchers and farmers, through their agricultural activities, have kept housing development off the Western landscape. As this land-use pattern collapses, trusts and enlightened units of governments are attempting to maintain open space and wildlife habitats. Thus far, these efforts have been laudable and success stories abound, but the task remains Herculean.

In many respects, the Colorado icon is both mirage and true terrain. The expanding success of land conservation groups in the Centennial State is heartening but certainly no cause for complacency. Colorado's history argues against the prospect of any but the most transitory ebbing of land development pressure. When things grow slack, promoters always find a new myth to build on.

It is entirely fitting that Buffalo Bill Cody, the original marketeer of the Western landscape, is buried on Lookout Mountain above Denver. The grave

There Lies a City

and adjacent museum are located 4 miles off I-70 next to the Pa-Hau-Ska Teepee Gift Shop. Legend and truth are both on display in the museum.

In 1867, Cody was hired as a buffalo hunter to provide meat for the construction crews working on the Kansas Pacific Railroad. He gained rapid acclaim for his ability to shoot the poor, shaggy beasts with a .50 caliber Springfield Rifle he called "Lucretia Borgia." A poem from that era helped create his nickname:

> *Never missed and never will,*
> *Always aims and shoots to kill,*
> *And the company pays his buffalo bill.*

The Wild West Shows began in 1872. Wild Bill Hickok, Annie Oakley, and Sitting Bull were among the legions of performers who traveled to New York, London, Rome, Paris, and Berlin to make spectacles of themselves. Reality and show business were badly blurred in Cody's mind. On July 17, 1876, three weeks after the Custer fight, he acted as a cavalry scout on a retaliatory raid against the Cheyenne. Cody killed and scalped Chief Yellow Hand and by October was back on the New York stage reenacting the murder for cheering mobs.

Cody was a far better showman than land developer. He invested in oil wells in Wyoming's Big Horn Basin which never met expectations. Cody tried a 2.5 million-acre real estate scam which went bust in Mexico. His mining company sold stock in the Campo Bonito gold, tungsten, and lead mine near Oracle, Arizona. No profit was ever made.

It seems his true calling was unrepentant prevarication—telling the big lie. In 1913, the Colonel William F. Cody Historical Pictures Company filmed a silent movie on the Pine Ridge Reservation in South Dakota. Cody hired poverty-stricken Sioux to re-create the Ghost Dance. Just 20 years before, this dance was used by Plains Indians in a desperate attempt to summon medicine enough to make all the white people disappear. Cody then peddled his movie in cities all over the country.

Buffalo Bill died in June 1917. There is a story told in Cody, Wyoming, that his wife sold his corpse to a group of Denver promoters for burial in a more marketable site. Even in death, Cody was shilling.

The grave, on Lookout Mountain just west of Denver, is surrounded by black and white stones and an iron fence. A marker above the grave calls him a "Noted Scout and Indian Fighter." The *Denver Post* on the day of his interment offered that "Never had a hero a more fitting resting place for his long sleep through eternity—never was a more appropriate site chosen for a grave" (June 17, 1917). The *Post* was right. For surrounding the burial area is a tangle of radio and television broadcasting towers and antennas. They seem like monuments erected to praise a media pioneer. The grave itself has Cody facing east toward Denver and the Front Range sprawl, with his back turned on the West. On the day he was buried, the *Denver Post* ran the following on page one:

> *General William F. Cody, "Buffalo Bill," rests in a tomb splendid. Human hands today placed him on the crest of Lookout Mountain on the dividing line between the wild and the tame. And such should be his sepulchre. For, was he not a bridge between Past and Future? There in the light of the setting sun are the memories of his achievements. There to the east, where the first gray of dawn starts the wheels of commerce in motion, there lies a city—a dream come true. (June 17, 1917)*

But across the Centennial State a different dream is beginning to manifest.

PART III
UTAH:
AN ELUSIVE ZION

^^^^^

12

THE
MORMON NATION

*When you have eliminated the impossible, whatever
remains, however improbable, must be the truth.*
—SHERLOCK HOLMES

Utah is a grand enigma. The place wears a veneer of familiarity in its Salt
Lake City skyscrapers, shopping malls, and well-maintained residential neighborhoods. People walking the streets appear no different than those seen in
Muncie or Spokane. Yet beyond architecture and fashion, beneath the patina
of its irrepressibly American cultural surface, Utah is a foreign nation.

The Church of Jesus Christ of Latter-Day Saints, also known as the LDS
or Mormon Church, is the single most powerful cultural and economic force
in Utah. To believers, the LDS President is considered a living Prophet who
serves as a dispenser of God's will, a model for personal aspirations, and
a fountain of wise homilies on correct ways of living. While Utah society
has grown more modern, the pronouncements of the Prophet are still
charged with extraordinary spiritual and secular power. His influence extends throughout Mormon society from individuals to corporations to government land management agencies.

While some 77 percent of Utah's 1.7 million residents are Mormon, 95 percent of Utah state legislators are LDS Church members. In addition, the
state's most powerful politicians are not merely Mormon, but are highly respected church leaders. Former Governor Norm Bangerter served as a Mormon bishop, and newly-elected Bob Bennett is active in the LDS Church.
Senator Orrin Hatch is the Gospel Doctrine Instructor for his LDS ward.
Although a non-Mormon woman was elected mayor of Salt Lake City in
1991, this historic breakdown of Mormon political dominance was due to a

∧∧∧∧∧

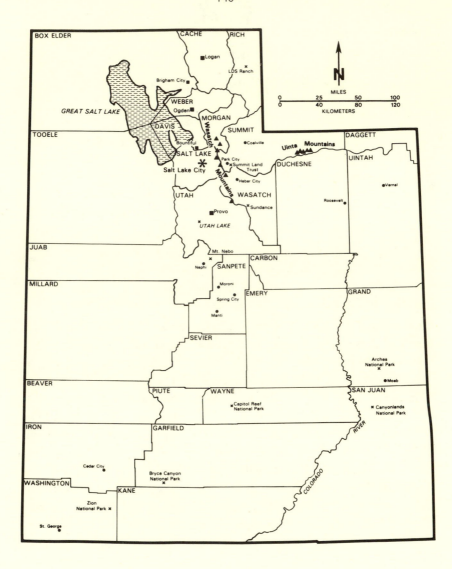

Map 6. Utah

Utah: An Elusive Zion

demographic shift within the city limits to a Gentile (non-Mormon) majority rather than a breakdown of Mormon voting ranks. Long-term residents of the state, Mormon and Gentile alike, still insist that nothing of any importance happens in Utah without the explicit or implicit approval of the LDS Church.

Today, 80 percent of Utah's population lives along the foot of the Wasatch Mountains in a continuous urban sprawl from Provo north through Salt Lake City to Ogden and Brigham City (Map 6). Life along the Wasatch Front is defined by rapid population gains, unchecked land development, mounting water shortages, rising outdoor recreational pressures, and tremendous air and water pollution problems. In many respects, the Wasatch Front and Colorado's Front Range face the same environmental dilemmas.

Yet the two places are only superficially alike. Colorado's land-use patterns and conservation attitudes have evolved detached from any overriding religious influence. Utah's landscape is not merely affected by a religion, it is a direct by-product of spiritual and cultural instruction. Utah did not happen by accident. It is the physical manifestation of the shifting Mormon search for the ideal, for the perfect place and way of life, for Zion on Earth. And like all such quests, the goal has proven to be elusive.

Solutions to Utah's many severe environmental problems have also largely eluded the state's conservationists and land-use planners. Only one land trust exists in the state, and it is located in the heavily Gentile resort community of Park City. Long-term Utah-watchers attest that tasks such as landscape conservation require much more than technical wizardry. Land and life in the Beehive State can only be understood by studying its history, the beliefs of the LDS Church, and the subtle nuances of Mormon culture. While Colorado is instantly comfortable to most of us, living in Utah feels alien—like a Peace Corps assignment or a dream where the world is just a half-turn askew. It must be explored as foreign terrain. Some of this knowledge can come from books, but often the most telling details emerge when we simply watch and listen.

MORMON CULTURE

Christmastime in Salt Lake City is a time of religious pilgrimage. Instead of Manger Square, Salt Lake City has Temple Square. Each December, the entire 10-acre walled-in block is decorated in a blaze of Christmas lights. Statues of the baby Jesus, assorted wise men, and camels are watched over by a single bright white star. On certain nights, the massed voices of the Mormon Tabernacle Choir spill out of the egg-shaped Tabernacle and fill the square with beautiful sound.

On such evenings, the two visitor centers are crowded with Mormons and Gentiles. Up a spiral ramp in the north visitors building is a massive white statue of Christ against a background of clouds, planets, and stars painted on the arching ceiling. Downstairs amid the Christmas trees and holiday merriment is a sign which reads simply "OBEDIENCE—The First Law." Outside young, expectant Mormon couples stand hand-in-hand gazing up at the 200-foot-tall west face of the holiest building in their world— the Salt Lake Temple. They speak no words. Their faces convey such innocent faith, such love for each other, that you turn away embarrassed to have intruded on such a private moment.

If you should accidentally bump into someone, he or she will politely apologize, smile with disarming genuineness, and pat you on the shoulder. These are not people carried away by the season. That same kindly look can be seen in small Utah towns on any summer afternoon.

It is easy to develop genuine fondness for the Mormon people. Most are trustworthy, good-hearted, and refreshing. If you had to lose your wallet, Utah would be the best place to do it. There's an exterior courtliness about this society which stands in stark contrast to the functioning of the corporate Mormon Church and certain intolerant features of Mormon secular life. The Salt Lake Temple, in the heart of Temple Square, lies at the vortex of both.

As with so many other things, Mormon Prophet Brigham Young chose the location of the Salt Lake City Temple. The structure was begun in 1853 when huge blocks of granite were hauled 20 miles from nearby Little Cottonwood Canyon to the construction site. Young initially wanted to build it of adobe, which was a very popular building material in Salt Lake at that time. He reasoned that through time adobe would be transformed into a harder, more durable material. Granite, he felt, had peaked and would only degrade and crumble. However, clearer heads prevailed, and the granite temple was completed in 1893, 16 years after Young's death.

The influence of astrology on the early Church can be seen all over the building in the carvings of the sun, the stars, and the moon in all its phases. A 12-1/2-foot golden statue of the Angel Moroni stands atop the central spire above the main entrance. The figure holds a long trumpet to his lips. Conservative Mormons believe that when Christ returns to Earth to usher in His

Millennial Kingdom, His first stop will be Salt Lake City. When this blessed event occurs, the statue of the Angel Moroni will spring to life and blow his trumpet to herald the beginning of God's rule on Earth.

The Salt Lake Temple, like all temples of the Church of Jesus Christ of Latter-Day Saints (LDS) and all Muslim mosques, faces east. However, this sacred structure also looks out on the twenty-story LDS office tower where the business of building up Zion is carried on. The elevation of the top of the LDS Administration Building and the State Capitol Building are officially listed as the same. The Mormon Church made a conscious decision not to overshadow the state's domed edifice of secular rule. However, from many angles, the church skyscraper looks more imposing.

Temple Square and Salt Lake City may be the heart of the Promised Land of Zion, but increasingly they are also the headquarters of the Church's sprawling corporate empire. The LDS office building has huge maps of the world carved into its south and north sides. Today, the Church of Jesus Christ of Latter-Day Saints claims to have 7.5 million members worldwide. While about 4 million Mormons reside in the United States, only 1.4 million of these live in Utah.

The LDS holy volume, the Book of Mormon, has been at least partially translated into eighty languages, including Tagalog, Pohnpeian, Trukese, Hebrew, Lingala, Afrikaans, Zulu, and Navajo. The Church boasts that 85 percent of the people in the world have some of the "B of M," as Saints often call it, available in their language. Some 36,000 LDS missionaries are at work in the world attempting to convert Gentiles to the faith.

The success of proselytizing has been uneven. There have been only 800 converts in India, but some 31,000 Tongans (29 percent of Tonga's population) have been baptized as a result of the zeal of young missionaries. To date, some ninety-seven countries and twenty-five territories, colonies, or possessions have organized LDS wards. Worldwide, there are about 1,800 stakes and 15,000 wards. A ward is equivalent to a congregation and contains around 400 to 450 members. A stake is analogous to a diocese. Each stake contains about 4 to 7 wards. Wards and stakes are used for regular services with temples reserved for family "sealings" and marriages. At certain times of the year, the Salt Lake Temple holds marriage ceremonies around the clock.

There are forty-three LDS temples in places as diverse as London, Johannesburg, Nuku'alofa (Tonga), Tahiti, Sydney, Taipei, Manila, Mexico

City, São Paulo, Santiago, Oahu, Los Angeles, Oakland, La Jolla (California), Cardston (Alberta), Washington, D.C., Seattle, Mesa (Arizona), Denver, and Toronto. In Utah, there are seven temples: Logan, Ogden, Jordan River, Manti, Provo, St. George, and Salt Lake. Within the dedicatory prayer at many temple openings, even those overseas, the Prophet or his representative has often referred to the place as Zion. As the Church has grown international, so has its concept of where God's Kingdom will be created.

Many of the 3.5 million foreign Mormons are nonwhites. The Church only allowed black males to enter the lay Aaronic priesthood in 1978. Women of any race are forbidden to do so. Given the rates of conversion in South America, the Philippines, and other regions, by the year 2020, the majority of Mormons may be nonwhites. This and the ban on women in the priesthood present the Church with some profound ethical issues to resolve.

The number of LDS Church members will rise dramatically if past trends continue. In the 1820s, church founder Joseph Smith and 5 others constituted the entire congregation. By 1847, when Brigham Young led the people to Utah, there were 35,000 Mormons. Thirty years later, at the time of Young's death, this had risen to 115,000. In 1890, when Prophet Wilford Woodruff denounced polygamy, membership stood at 188,000. In 1920, there were 526,000 members and by 1950, this had more than doubled to 1.1 million. After 1950, a numerical explosion occurred. This is the demographic litany: 1960—1.7 million, 1970—2.9 million, 1975—3.6 million, 1980—4.7 million, 1985—5.9 million, and by the end of 1992, there were at least 7.5 million Mormons in the world.

How fully "Mormon" many of these converts are is unknown. This is particularly true overseas where religious conversion, whether Catholic or Mormon, is often a matter of baptism followed by adoption of a veneer of the new faith superimposed on or interwoven with indigenous beliefs. However, current estimates show a worldwide membership of 10 to 12 million by the year 2000, with forty-three additional temples built.

Despite this international growth, walking through Temple Square a few days before Christmas has the feel of a small-town gathering of members of a local, nonuniversalizing congregation. The square is also a symbol of the trials and struggles of the early Church and the success of the people over unrelenting oppression. Today, Mormonism has little to do with utopian visions of a bucolic agricultural landscape or even the building up of a single city for God. The current corporate Church is defined by the quest for converts, political power, social control, and monetary profit.

Utah: An Elusive Zion

CONTROL OF KNOWLEDGE

In the Bible, Daniel tells the King of Babylonia that great nations will be destroyed by an unnamed powerful force originating from a high mountain. Many Mormons believe this "mountain" refers to the LDS Church, which came to fruition deep in the Rockies. To them, the Church is mandated to accumulate wealth and influence because we are living in the "last days" before the Millennium, when Christ and His chosen people, the Mormons, will rule the entire world. The third Mormon Prophet, John Taylor, once said:

> *The Constitution of the United States was not a perfect document, it was one of those stepping stones to future development [with] a higher law, more noble principles, ideas that are more elevated and expansive, [a kingdom] disunited from all forms of government, . . . a kingdom separate with a separate form of government [that] shall be given to the Saints. We seek to carry out the designs of the Almighty and his representatives upon the earth. And if we do these things, in the name of Israel's God, we shall arise and flourish, and Zion will become a terror to all nations. (*Journal of Discourses, LDS Church, 7: 123 – 125*)

The control of information is a central element in this social control. In 1972, Leonard J. Arrington, a respected historian, was appointed director of the LDS Church Historical Department. Arrington, a Mormon, had long wished his church would make available historical documents which for decades had been locked up in a granite cave, known as "The Vault," in Little Cottonwood Canyon. He would say, "It is unfortunate for the cause of Mormon history that the Church Historians' Library, which is in possession of virtually all of the diaries of leading Mormons, has not seen fit to publish these diaries or to permit qualified historians to use them without restriction" (Naifeh and Smith, 90).

Over the years, Arrington attempted to open Mormon history to scholarly analysis. In 1981, Elder Boyd Packer gave a speech in which he expressed the Church's discomfort with scholarly research and, indirectly, Arrington.

> *I have come to believe that it is the tendency for many members of the Church who spend a great deal of time in academic research to begin to judge the Church, its doctrine, organization, and leadership, present and past, by principles of their own profession. . . . There is a temptation*

*for the writer or the teacher of Church history to want to tell everything,
whether it is worthy or faith promoting or no. Some things which are
true are not very useful. . . . Be careful that you build faith rather than
destroy it. . . . A destroyer of faith—particularly one within the Church,
and more particularly one who is employed specifically to build faith—
places himself in great spiritual jeopardy. He is serving the wrong master and unless he repents, he will not be among the faithful in the
eternities. . . . In the Church we are not neutral. We are one-sided.
There is a war going on, and we are engaged in it. . . . There is a limit
to patience of the Lord with respect to those who are under covenant to
bless and protect His Church and kingdom upon the earth but do not
do it. (Naifeh and Smith, 137)*

Shortly thereafter, Arrington, the well-respected dean of Mormon historians, was fired from his position. The so-called "Arrington Spring" was
over. He was replaced by G. Homer Durham, a conservative with no background in history.

The Church's concern about historical documents eventually resulted in
the bizarre "White Salamander" episode. During the mid-1980s, the Church
made numerous payoffs to a master forger named Mark Hofman. Hofman
claimed to have found an 1830 letter written by a church founder which
stated that Joseph Smith had been told of the gold tablets containing the
Book of Mormon by a white salamander and not by the Angel Moroni, as
church history had long said. The satanic overtones of this alleged letter
caused a panicked church hierarchy to buy Hofman's silence. In a desperate
attempt to keep LDS checks coming, Mark Hofman murdered two people
and nearly killed himself when a bomb he was carrying in his car exploded
in downtown Salt Lake.

FINANCIAL POWER

The modern Mormon Church has accumulated a diverse and sophisticated
array of tightly controlled corporations and a bulging stock portfolio managed by faithful brokers quietly staring at computer screens.

The investment strategy of the LDS Church also mirrors concern for the
management of information. The Bonneville International Corporation is
the Church's broadcasting subsidiary. Through this media conglomerate, the

Church operates television stations KBYU in Provo, KIRO in Seattle, and KSL in Salt Lake. KSL is transmitted by 121 translator relay stations around the Rocky Mountain West. The Federal Communications Commission has concluded that "KSL-TV possesses the largest Area of Dominant Influence (ADI) of any station in the Continental United States" (*FCC Reports*, 255). Bonneville International also owns twelve commercial radio stations, including WRFM in New York, KBIG-FM in Los Angeles, WCLR-FM in Chicago, and KOIT-AM/FM in San Francisco. Nearly every LDS ward house has a satellite dish on the grounds. This is not so they can pick up ESPN. Bonneville Satellite, a subsidiary of Bonneville International, beams church programming to members around the world.

Despite pronouncements by the LDS General Authorities about the evils of modern broadcasting, the Church has a long history of associating with the media. In the 1940s, Darryl F. Zannuck and Twentieth Century Fox, with cooperation from the Mormon Church, produced the heroic and romantic motion picture, *Brigham Young*. Dean Jagger played Young. Supporting players included Tyrone Power, Linda Darnell, and Mary Astor. Horror film legend Vincent Price was cast as Joseph Smith.

In the realm of print media, the LDS Church owns 300,000 shares in the Times-Mirror Corporation, making the Church the second largest stockholder. Times-Mirror owns numerous newspapers, such as *The Los Angeles Times*, *Newsday*, *The Dallas Times Herald*, *The Hartford Courant*, *The Denver Post*, and *The Sporting News*. The Church also founded and operates the immensely profitable *Deseret News* and Deseret Books, both in Salt Lake. The total value of all church holdings in the broadcast and print media is over $550 million.

The Mormon Church possesses a portfolio of public utility bonds and common stocks worth over $255 million. It holds title to numerous commercial properties in Salt Lake City, such as the ZCMI Center Mall, the Kennecott Office Building, the Hotel Utah, Temple Square Hotel, Eagle Gates Apartments, and the Promised Valley Playhouse, which are valued at some $204 million. In addition to being the largest private landowner in Utah, the LDS Church owns some 30,000 acres of other commercial property located in places as far afield as Los Angeles, Boston, Kansas City, Honolulu, Tokyo, São Paulo, Seoul, and Sydney. The cumulative (1983) value of all such commercial holdings has been conservatively estimated at $757 million. Agricultural lands possessed by the Church total some 928,600 acres with an unknown appraised value. The 316,000-acre Deseret Ranch in Florida and the 201,000-acre Deseret Land and Livestock Company in northeast Utah are the single largest ranches in those states.

The amount of tithing by Mormons is difficult to establish. Estimates range up to $2 billion per year. Maintaining an international network of wards and operating the Church's many businesses are certainly expensive. However, because church officials no longer publish an annual financial report, yearly expenses and revenues cannot be calculated precisely. In their painstakingly researched book, *The Mormon Corporate Empire*, John Heinerman and Anson Shupe estimated the 1983 value of all LDS Church assets at $7.9 billion. A long-time Mormon-watcher in Salt Lake believes this figure is now "probably two to five times too low."

There is also a grey area between actual church wealth and that of its members. The Church often merely has to suggest that wealthy Mormons step forward to finance a project, and they will. Mormon J. Willard Marriott, founder of the Marriott Hotel chain, has donated several million dollars' worth of stock to the Church. Therefore, when this sort of economic access is considered, the true financial power of the Mormon corporation dwarfs the $7.9 billion figure. For well-connected Mormons, Utah is the promised place, the land of silk and money.

LIFE IN ZION

The central place of Utah, Salt Lake City, is a very liveable and safe—yet extremely peculiar—community. The stereotype is that it is boring. Nothing could be further from the truth. The complex cultural undercurrents manifest in many aspects of day-to-day life.

After living in Salt Lake for a few months, you begin to hear curious conversations at the local Smith's supermarket, at the University of Utah, and at bookstores around the city. While more than half the residents of Salt Lake proper are Gentiles, it is clear that they perceive themselves as an ethnic minority. Gentiles critical of the Church often compare the hierarchy of Mormon General Authorities to the old Kremlin bureaucracy. In their view, the President of the Church has the same power as the Premier, the Twelve Apostles serve as the Supreme Soviet, and the Quorum of the Seventy are the Politburo. Gentiles widely believe that if you're not a Mormon in Utah, your upward mobility in government and many businesses is severely limited. When Russian poet Valentina Inozemtseva visited Salt Lake City in

1987, she was asked what the place reminded her of. "The Soviet Union," was her terse reply.

One Halloween when I was living just off 13th East, only a handful of trick-or-treaters from the neighborhood came to the door. Certain children even seemed reluctant to speak, even when greeted on the street. Friends finally explained that some Mormon families instruct their children not to socialize with Gentiles on the block and definitely not to trick-or-treat at their homes. The teenage daughter of Salt Lake friends said the division between "Mo's," as she called Mormons, and Gentiles was particularly strong in her high school. "In my school," she said, "the Mo's run everything and gain support from the Mormon teachers and administrators in doing it. I'm going to the U [the mixed University of Utah]," she continued, "While all the serious Mo's are going to the Y [Brigham Young University]. I can't wait to graduate."

The essential eccentricity of Salt Lake turns up everywhere. On hot July afternoons, the city's immensely popular Snelgroves ice cream parlors are packed. Despite the heat, some women wear full-length Mormon "garments" beneath their clothes. These union suits are worn by true believers in preparation for the Millennium, when all Gentiles will be struck naked by the Lord, and only Mormons will be spared the embarrassment.

Cultural distinctiveness materializes even in local toy stores. Two of the most popular board games in Salt Lake are "Treasure Trek: A Book of Mormon Adventure Game" and "Celestial Pursuit: The Ultimate in LDS Trivia." A church publication, *Family Home Evening Resource Book*, also contains Mormon games which reinforce the religious doctrine children should be learning.

The Church's optimism regarding life in the city is also revealed by the weather forecasters on the church-owned KSL television station. With broad smiles, they often refer to the city's choking winter smog as "haze" or "fog." The Zero Population Growth organization has rated the environmental quality of Salt Lake City 191st out of 192 U.S. cities, in part because of its severe air pollution problem.

Salt Lake is often described by Gentiles as "The Emerald City," where the President of the Mormon Church is the great and powerful Oz. Nowhere else in America is a religious leader so dominant and such a singular, Millennial culture so in charge.

When the English geographer Sir Richard Burton passed through the city in 1860, he told LDS leaders their religion was a creative admixture of

Old Testament bluster, Jewish mysticism, transcendentalism, Millennialism, Freemasonry, and Islam. To this they simply smiled. Today, upon visiting Temple Square and talking to Mormon counselors, their response to questions about their theology will often be simply: "This is the truth—just cast aside doubt."

CONCLUSION

Utah is a singular nation rife with contradictions and rich in opportunities to amaze. There is much to admire and be puzzled by along the urban Wasatch Front and in the suburbanizing back-valleys east of the mountains.

Salt Lake City is the cradle of Mormon settlement. Its mountain watersheds, valleys, and lakeside environments are under the most intense development pressures in the state. Much of this growth is fueled by distant water funneled into the city as part of the immense Central Utah Project. Provo is the state's "Happy Valley," a place of industrialization and tremendous population growth which, for now, retains a substantial agricultural land base and superb wildlife habitats. North of Salt Lake City, Bountiful, Ogden, and Brigham City manifest intriguing variations in the interplay between land conservation and development in a suburban corridor dominated by "high tech" and defense industries.

The formerly agrarian Wasatch back-valleys are the new stage for Utah's population growth and landscape change. Wasatch County is being transformed by reservoirs constructed to complete the Central Utah Project, with only parts of it saved by conservation efforts such as The Nature Conservancy's Strawberry River Project. Summit County is wonderfully eclectic. It is traversed by the Mormon Trail, the home of polygamous Mormon clans, the scene of big game and raptor preserves, and the location of Park City, one of the West's most popular destination area ski resort communities. And finally, even Sanpete County, Utah's agrarian heartland, is confronting increasing recreational development and environmental disturbance. Sanpete is the last bastion of the traditional "Mormon landscape" and stands as a superb candidate for the creation of a new kind of *cultural* national park— one which celebrates the original Mormon role as enlightened stewards of a fine land.

These landscapes are an encyclopedia of environmental and cultural history. They are brimming with the geography of Mormon dreams. Questions abound. How did such a remarkable natural landscape become so environmentally changed since the original Mormon colonization of 1847? Importantly, is this what Mormon culture has been striving to create, or is it an unintended side effect of their quest for Zion? And finally, is anything being done to conserve lands of great ecological and cultural importance in the core of the Mormon world?

The answers to these questions do not begin in Utah. To properly understand how these Mormon places came to be, we must travel back to the beginning, to the 1820s in the Finger Lakes District of upstate New York and into the mind of a very unusual young boy.

13
MONEY-DIGGING AND SEER STONES

In the early nineteenth century, America was in the throes of a religious revival known as the Great Awakening. Socialistic, communal, and utopian sects, such as the Shakers and Owenites, were becoming widespread. Camp revival meetings and traveling preachers served as summertime entertainment for rural folk. Palmyra, in New York's Finger Lakes District, was one of the popular stopping points for circuit-riding evangelicals.

In Palmyra, a once-bedridden, introspective young farm boy grew confused but fascinated by the conflicting claims of religious orators who passed through town. He wondered which of these faiths was the true church—the real representative of God's will. In 1820, the fourteen-year-old walked into the maple woods to pray for guidance. At once he was "seized upon by some power" and saw a "pillar of light exactly over my head" within which two glowing personages seemed to float. He later recounted, "one of them spoke unto me, calling me by name and said 'this is my beloved son, hear Him'" (Smith, 5–7). Hurrying to town the boy excitedly described his eerie encounter. No one believed him. He felt persecuted.

In the 1820s, farming was not a profession for a dreamer. It required unrelenting labor and sacrifice. The boy hated tilling the soil as only a farmer's son can and gazed into a mortgaged future. The vision, he knew, was his salvation from a life of drudgery and boring sameness.

The boy immediately took up digging for "treasures" of gold and money which he imagined lay buried in the hills south of the Erie Canal. Divining rods, seer stones, incantations, magic wands, crystals, and stuffed toads were

∧∧∧∧∧

used to attempt to locate mysterious underground stores of wealth. Immersed in magical arts, the boy naturally gravitated to an itinerant fortune teller named Walters, who showed up to ply his trade in Palmyra. Walters claimed to have discovered an ancient Indian account telling of hidden treasures. As Walters spun stories linking his "manuscript" to what we now know are Indian burial mounds, the boy sat enthralled. Walters read to the locals from his "ancient record" in what sounded like a long-dead Indian dialect. Some would later claim it was simply Cicero in Latin.

After Walters moved on, the young man resumed his independent geomancy. Once he told a neighbor a trunk full of gold watches lay beneath his pasture. Another time he told a farmer that a large sum of money was buried on his property. The young man told them that to assure the money did not magically move, a black sheep must first be walked around the spot with its throat cut to appease the dangerous ghosts which protected the treasure. When the digging proved fruitless, the young man solemnly intoned that they had failed to perform the ritual correctly. He would later claim that while digging a well, he found a "seer stone" through which he could see ghosts, infernal spirits, and mountains of gold. One of his supposed treasure sites was a large hill, near his father's farm, which later would yield wealth far grander than even he could imagine.

On the evening of September 21, 1823, the young man had a second even more detailed and enrapturing vision of a personage. He later recalled:

> *He had on a loose robe of most exquisite whiteness. . . . He called me by name and said unto me that he was a messenger sent from the presence of God to me and that his name was Moroni. . . . He said that there was a book deposited, written upon gold plates, giving an account of the former inhabitants of this continent and the sources from which they sprang. He also said that the fullness of the everlasting gospel was contained in it, as delivered by the Savior to the ancient inhabitants; also that there were two stones in silver bows—and these stones, fastened to a breastplate, constituted what is known as the Urim and Thummim—deposited with the plates; and the possession and use of these stones were what constituted "seers" in ancient and former times; and that God had prepared them for the purpose of translating the book.*
> (Book of Mormon)

The personage forbade the young man from immediately digging up the gold plates, instructing him to wait until he was deemed worthy.

Money-Digging and Seer Stones

His attempts at money-digging and conjuring then grew even more adventurous. In 1826, when he was twenty, the young man was put on trial in Palmyra for being "a disorderly person and an imposter" (Brodie, 18). Fifty-one of his neighbors signed an affidavit which described him as "destitute of moral character and addicted to vicious habits" (Brodie, 18). He acknowledged he had a seer stone which he "occasionally looked at to determine where hidden treasures in the bowels of the earth were" (Brodie, 30). After the public scolding he received at the trial, the young man, Joseph Smith, backed off overt magic and began to focus his abundant imagination on his life's work.

The search for gold, a gnawing confusion over religion, and the mound-building of paleo-Indians formed three central components of Joseph Smith's psyche. Smith read in the *Palmyra Reflector* that during the excavation of the Erie Canal, workmen found several brass plates along with skeletons and pottery fragments. Smith began to invent theories about the Indian mounds, even imagining details about Indian clothing, cities, religion, and ways of making war. In 1823, a Vermont preacher named Ethan Smith published a book called *View of the Hebrews; or the Ten Tribes of Israel in America*. The volume was a compilation of folk and scholarly belief that Native Americans were remnants of the Lost Tribes. Cotton Mather and William Penn were among the many adherents to this intriguing notion. Given Joseph Smith's interests, he was no doubt aware of *View of the Hebrews*, or at least the theory it espoused.

THE BOOK OF MORMON

In 1827, Smith announced that he had at last received permission to dig up the gold plates foretold by the Angel Moroni. The plates lay, he said, on his father's place in a large hill which had the look of a massive Indian mound. It was actually a glacial drumlin deposit which Mormons today call Hill Cumorah. Smith secreted the plates in his home and warned his wife, Emma, that she would die instantly if she set eyes on them.

The plates were said to be densely inscribed in a mysterious "Reformed Egyptian" script. Smith performed his "translation" of the "text" from behind a curtain, using the "Urim and Thummim," as he dictated to one of several people who acted as recorders. Eleven witnesses signed testimonials that they had seen the golden plates, but some later recanted their claims. No outside evidence has ever come forward to prove that the plates existed.

Utah: An Elusive Zion

Nonetheless, Smith's "translation," known as the Book of Mormon, was first published in 1830 and had immediate impact.

The Book of Mormon is a saga of three races locked in New World warfare. In the Book, a young Hebrew prophet named Nephi left Jerusalem in 600 B.C. and sailed to America with his father Lehi and a group of followers. Once they landed, the Hebrews split into three groups: the Nephites, the Jaredites, and the Lamanites. The Nephites, a race of fair farmers and temple builders, spread across North and South America only to be attacked by a group of "savage and bloodthirsty" hunter-gatherers known as the Lamanites. The Jaredites were swept into the fray. To Joseph Smith, the Indian mounds of upstate New York were the ruins of forts where the Nephites were finally wiped out. The Jaredites also succumbed.

Smith believed a Nephite prophet named Mormon (the father of Moroni), had desperately written down the history of this war on the gold plates in the hope that a future prophet would find the true way to God. To Smith, the Lamanites triumphed only to be turned dark-skinned by God as a mark of His displeasure. For Mormons, Lamanites were the ancestors of the Indians of North and South America, Polynesia, Micronesia, and other regions.

Joseph Smith recalled the line in the Bible where Jesus says, "Other sheep I have, which are not of this fold, them also I must bring, and they shall hear my voice." He took this quite literally. Smith believed that the white figure of Quetzalcoatl in Aztec mythology was simply Jesus ministering in the New World after His death and resurrection. It was a brilliant and appealing idea—Christ in America!

To Mormons, first there was the Old Testament, then the New Testament, and finally the Book of Mormon, which is subtitled "Another Testament of Jesus Christ." Smith had woven together all the threads of his turbulent life into one marvelous, all-embracing story. In 1976, Mormon historian Fawn Brodie wrote: "Thus where *Views of the Hebrews* was just bad scholarship, the Book of Mormon was highly original and imaginative fiction" (Brodie, 12). Ms. Brodie was promptly excommunicated.

ZION

The concept of Zion has always been the core of Mormon life. Yet perhaps no other theological and cultural idea in Latter-Day Saints (LDS) reality has changed as often and in such a complex manner. Early Mormons sought to prepare communally for the Millennial rule of God on Earth by creating a society, a place, a state of grace known as Zion. The landscape was perceived as an arena where their faith and willingness to work would be tested. If they strove righteously, God would reward them with a productive, safe homeland. But to earn this promised land, the people first must have their wills tempered by oppression and their patience tried by enduring an exodus across a vast wilderness into a difficult land.

For early Mormons, the central question of life was "Where is our rightful place, our Zion?" By 1830, the membership of the Church of Jesus Christ of Latter-Day Saints had grown to only 280. However, the dream was to strike off in search of a divinely guided fresh start in a new land. Smith wrote vaguely of the geography of Zion:

> *And now behold I say unto you, that it is not revealed and no man knoweth where the City of Zion shall be built, but it shall be given hereafter; I say unto you that it shall be on the border of the Lamanites.* (*Doctrine and Covenants*, LDS Church, 28:9)

Given that much of North America was then Indian land, Smith was undecided about where to go and chose to bide his time while building church membership through lectures, prayer meetings, and distribution of the Book of Mormon.

EXODUS

The Mormon exodus began in 1831 when hundreds of Saints moved from New York and New England to Kirtland, Ohio, near Cleveland. A temple was constructed in 1836, and Joseph Smith pronounced the place "Zion." However, armed conflicts arose with Gentiles over polygamy and economic competition. Locals also feared the Mormons were trying to enlist the support of Indians, whom the LDS Church viewed as Lamanites, and therefore, fellow Israelites.

In 1837, Smith sent Heber Kimball and other confidants to England to gain converts and monetary support. By 1838, the Mormons had moved on to Independence, Missouri, which Joseph Smith variously described as the final gathering place, Zion, the City on the Hill, and the New Jerusalem. It was *here*, he said, the people would build an eternal city and await the divine advent. However, violence again broke out between Mormons and Gentiles.

In response, the Mormons organized the "Danites," or "Destroying Angels," as a military force. Governor Boggs of Missouri grew incensed at this "terrorism" and issued an order to "exterminate or drive the Mormons from the state." Outnumbered, the Saints grew downcast and made preparations again to move on. On a cloudy spring morning, long caravans of Mormon wagons pulled dejectedly out of Independence.

Joseph Smith, the President and Prophet of the Church, promptly announced a face-saving revelation that Independence would be Zion's spiritual center and that they would one day return in triumph. In a brilliant, bold stroke, Smith also proclaimed all of North and South America would be Zion as well. This audacious revelation served two important purposes. First, it made the Saints more optimistic as they were once again chased from their homes. Second, it expanded and generalized Zion, which, up until then, had been particularized at specific sites.

This was the beginning of an important pattern. The concept of Zion could be concrete or abstract, local or continental, as circumstances demanded. This flexibility saved the Church or altered its direction on many occasions. Faced with extinction, the church leadership had learned an invaluable lesson—sometimes it is necessary to revise history and doctrine to secure the obedience of the people.

In 1839, the Mormons settled beside the Mississippi River in the town of Nauvoo, Illinois. Within two years, the community had 15,000 residents and was the second largest city in the state. As the Mormons prospered, discord with Gentiles once again grew violent. Porter Rockwell, Smith's personal bodyguard and renowned Destroying Angel, was jailed for returning to Missouri and seriously wounding Governor Boggs. Blood atonement was much more than a theological principle in the early days of Mormonism. An "eye for an eye" meant exactly that.

In 1842, Joseph Smith joined the Freemasons and became fascinated by their rituals. "The Craft" is said to date back to the building of King Solomon's Temple. Smith wanted temples, too. Six weeks after being installed as a Mason, he gathered the upper echelon of the Church for instruction "in the principles and order of the priesthood, attending to washings, anointings, endowments, and communication of keys" (Smith, 5:2). Smith drew heavily from Masonic rituals in elaborating the temple ordinances which later became the "endowments" of the LDS Church.

In the temple, people were to be stripped, washed, and anointed and then, as is true in Masonic ceremonies, given a temple "garment" to wear. In matters of dress, handgrips, keys, oaths, and secret passwords, the similarity between Mormonism and Freemasonry is striking. Even the Mormon beehive symbol, which epitomizes an industrious, obedient society, is drawn from the Freemasons. Smith reveled in the pageantry. Fawn Brodie wrote, "This is why Joseph could derive enormous satisfaction from playing the role of God in the temple allegory" (Brodie, 283).

By 1844, Smith had selected a Council of Fifty, which crowned him "King of the Kingdom of God," and announced his candidacy for the presidency of the United States. Gentile mob actions and armed skirmishes grew more frequent. In retaliation, Smith ordered destruction of the presses of the reform-oriented newspaper *The Nauvoo Expositor*. Joseph and his brother Hyrum were caught fleeing the city and were jailed in nearby Carthage to await trial. A mob of "militiamen" gathered outside the jail and soon dragged both men from their cells and murdered them. The Prophet, President, and founder of the LDS Church was gone, and a pall of doubt and fear descended over Mormonism like a freezing fog.

The succession to the Mormon presidency resulted in a deep schism among the people. Those following Joseph Smith's son split off to form the Reorganized Church of Jesus Christ of Latter-Day Saints (RLDS), which today has about 500,000 members, mostly in the Midwest. However, the main corpus of the Church followed Vermonter Brigham Young, who, as President of the Twelve Apostles, was heir to the position of infallible living Prophet.

Unlike Smith, Young was less a visionary theologian than a hard-nosed organizer and businessman. He knew that as long as Mormons settled among Gentiles, the pattern of violence would persist. Young was aware of the geography of the Great Basin from the reports of Jedidiah Smith and John Fremont. While others in the Mormon hierarchy considered Texas, Chihuahua, California, and Vancouver Island excellent places to build a city for God, Young decided that the emptiness of the Great Basin and Wasatch Mountains of present-day Utah was ideal. No one else seemed to want it. Even the region's Gosutes and Utes seldom camped there.

In 1846, the United States was at war with Mexico. The Mormon trek west continued with the 500-member Mormon Battalion splitting off as part of General Stephen Watts Kearney's Army of the West. Their pay provided needed cash for the main colonizing body which was bivouacked in Winter Quarters, Nebraska, on its way to the promised land of Utah.

Some 600 people in Nebraska died from cholera and other diseases that winter. Brigham Young tried to counteract the consequences of these deaths in three intriguing ways still manifest today. First, there was a call for proxy baptisms of the dead, so they could enter Celestial Glory (also known as Zion) or even Exultation, the highest of the three rungs within this heaven. Second, there was renewed emphasis on plural marriage, so all female Saints could bear children to repopulate the Church. Lastly, a renewed call was made for missionary work in the British Isles, Scandinavia, and Polynesia.

To Young, the answer was an isolated homeland, safety in numbers, and the accumulation of wealth. Guided by these basic principles, the Mormon people once again rose to their feet and headed west.

14

COLONIZATION AND
THE MORMON LANDSCAPE

THE SALT LAKE VALLEY: THE PROMISED ZION

The Mormon exodus to Utah largely paralleled the Oregon Trail on the north bank of the Platte River. This path, designed to avoid mixing with Gentiles, became known as the Mormon Trail. On July 24, 1847, the Mormon Pioneer Company, consisting of 148 people, crossed the Wasatch Mountains of present-day Utah and descended oak-mantled Emigration Canyon. Brigham Young, weakened by tick fever, looked out on the valley's stream-laced benchlands and from a reclining position announced, "It is enough. This is the right place."

To Young, God's promised gift was this strange and beautiful landscape, which would test the people's faith. While Joseph Smith had spoken of the Saints as stewards obliged to a divine landlord, the ever-pragmatic Brigham Young said:

> *The work of building up Zion is in every sense a practical one; it is not mere theory. To possess an inheritance in Zion only in theory—only in imagination—would be the same as having no inheritance at all. It is necessary to get a deed of it, to make this inheritance practical, substantial, and profitable.* (*Journal of Discourses*, LDS Church, 9:284)

Young was blunt: "My policy is to get rich. I am a miser in eternal things. Do I want to become rich in things of the earth? Yes, if the Lord wishes" (*Journal of Discourses*, LDS Church, 2:144). He exhorted Mormons

∧∧∧∧∧

to "make our own heaven, our own paradise, our own Zion" (*Journal of Discourses*, LDS Church, 9:170) and reminded them, "I have Zion in view constantly for we are not going to wait for angels, or for Enoch and his company to come and build up Zion, but *we* are going to build it" (*Journal of Discourses*, LDS Church, 9:284).

Upon arriving, some Saints were not at all enraptured by the place. Lorenzo Young wrote in his diary:

> *This day we arrived in the valley of the Great Salt Lake. My feelings were such as I cannot describe. Everything looked gloomy and I felt heartsick. (in Little, 98)*

His wife, Harriet Snow, offered:

> *We have travelled fifteen hundred miles to get here, and I would willingly travel a thousand miles farther to get where it looked as though white men could live. (in Little, 98)*

However, many others were enamored of the new Zion. Mormon pioneer Thomas Bullock wrote:

> *I could not help shouting, "hurra, hurra, hurra, there's my home at last." The sky is very clear, the air delightful, and all together it looks glorious. (Bullock, 21)*

A recurrent Mormon myth is that the pioneering Saints were like Israelites wandering through an unknown wilderness guided only by God. In truth, Lansford Hastings's emigrants' guidebook and John Fremont's book on his exploration of the Rockies were well studied by church authorities. Portions of those works had been published in the Nauvoo newspapers and provided the Saints with information on the soils and climate of their eventual home.

It has also been widely claimed that during settlement the Salt Lake Valley was, as Prophet George A. Young called it in 1865, "a howling desert . . . in the heart of the Great American Desert" (*Journal of Discourses*, LDS Church, 2:177). Mormons were supposed to have made the desert "bloom as the rose."

In fact, most settlers in 1847 saw the valley as a highly favorable terrain (Map 7). Over and over in their diaries, pioneers noted streams, flood-

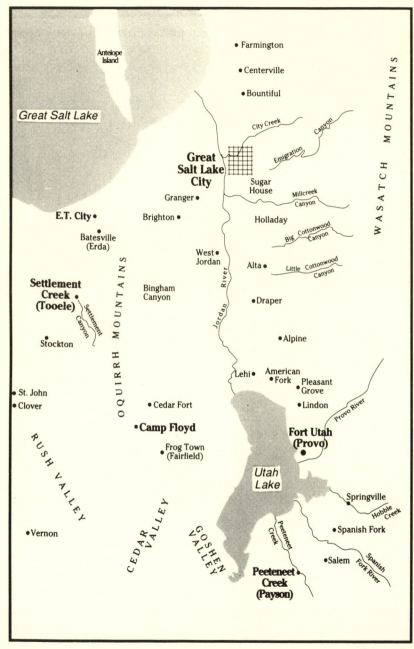

Map 7. Great Salt Lake Region, 1847–1869 (redrawn from Campbell, p. 55)

Utah: An Elusive Zion

plains, excellent soils, tall grass, and a dry climate tempered by cooling canyon winds. Few even mentioned the word desert. Future Prophet Wilford Woodruff wrote:

> *We came in full view of the valley of the Great Salt Lake—the land of Promise, held in reserve by the hand of God as a resting place for the Saints. We gazed with wonder and admiration upon the most fertile valley spread out before us for about twenty-five miles in length and sixteen miles in width, clothed with a heavy garment of vegetation and in the midst of which glistened the waters of the Great Salt Lake, with mountains all around towering to the skies, and streams, rivulets, and creeks of pure water running through a beautiful valley. . . . To gaze upon a valley of such a vast extent surrounded with a perfect chain of everlasting mountains covered with eternal snow, with their innumerable peaks like pyramids towering towards heaven, presented at one view to us the grandest scenery and prospect that we could have obtained on earth. (Woodruff, 313)*

Mormons also once claimed to have invented irrigated farming out of necessity. However, several years prior to their arrival in Utah, Apostle Orson Hyde had been dispatched to tour Israel, Egypt, Lebanon, and Syria, in part to study irrigated agriculture. His explorations of the Nile delta would have provided extensive exposure to the workings of irrigation. In addition, Mormon traders and members of the Mormon Battalion traveled through New Mexico, Arizona, and California where Hohokam Indian and Spanish-influenced irrigation techniques were observed.

Upon arrival in the Salt Lake Valley, the Mormons immediately built a diversion dam on City Creek and within eight days had planted seed potatoes in a 53-acre field watered by irrigation ditches. If the Saints invented irrigated agriculture in Utah, it was indeed a spontaneous technological revelation.

EARLY SETTLEMENT

In 1847, Utah was nominally a part of Mexico. President Polk had given his approval to the Mormon colonization effort, since it would help shore up U.S. claims to the area. Without legal authority other than Polk's recognition, the Mormons devised their own land tenure system.

The design for the City of God came from Joseph Smith and his Plat for the City of Zion. This, the Imitatio Dei, or sacred model of Mormonism, necessitated centralized land-use planning. The design of its regular grid of streets and blocks was four-square with the cardinal directions of the compass.

"Plat A," in what was then called Great Salt Lake City, consisted of one square mile with a 40-acre temple block at its center. Each 10-acre block was subdivided into eight lots 10 by 20 rods in size. Streets were an incredible 8 rods wide to allow ample room for turning ox-driven wagons. All houses were to be built in the middle of each lot and set back from the street 25 feet to inhibit the spread of fire. Four squares were held back for public use as schools or parks. Plat A, containing 114 blocks, was surveyed within a month of initial settlement.

Brigham Young served as the final arbiter in land planning. He personally claimed "two tiers of blocks running north and south through the city, and each of the apostles made reservations as well" (Fox, 42). The so-called Big Field, lying southeast of the city, was set aside for agriculture.

When 863 applications were received for a total of 11,005 acres, Young decided only 5- and 10-acre tracts should be surveyed away from the city. The Church tightly controlled land subdivision in order to prevent speculation. Brigham Young announced that "no man should buy any . . . but every man should have his land measured off to him for city [purposes] and for farming purposes what he could till" (Linford, 127).

Equitable land dispersal was attempted in the fashion described by Joseph Smith as consecration and stewardship. However, many Saints remembered that in Nauvoo, the High Council had bought land for $37–$100 per acre, platted a city for God, and sold three-quarter acre lots for an average of $500 each. To some, this was blatant land speculation. The fact that in Salt Lake City, the Church's General Authorities received large parcels of land while common church members had to apply and wait to get a small tract did nothing to allay fears. In 1848, Plat B was surveyed to accommodate the arrival of hundreds more church members. The expansion had begun.

Mormon historian Leonard Arrington's extraordinary book *Great Basin Kingdom* chronicles the communal character of Mormon settlement and economic life. The communal model largely holds true for irrigation companies minded by water masters and for the United Order agricultural and manufacturing cooperatives headed by Mormon bishops. The Church also initially adopted a communal approach to exploitation of resources in the Wasatch Range. In 1847, Brigham Young declared: "There shall be no private ownership of streams that come out of the canyons, nor [of] the timber that grows on the hills. These belong to the people—all the people" (Roberts, 3:269). He issued an epistle on September 9, 1847, which mandated that only dry wood unsuitable for timber be used as firewood and that big-toothed maple shrubs be left for sugaring.

Mormon ideas about stewardship and the creation of a Zion on Earth combined with communalism and tithing to present the appearance of a people prepared to use natural resources with more wisdom than other Western settlers. Mormon theology, while emphasizing economic growth, proclaimed that the betterment of people and nature would culminate in a "peaceful kingdom" prior to the Millennium. Even Brigham Young said:

> Let the people be holy and the earth under their feet be holy. Let every animal and creeping thing be filled with peace; the soil of earth will bring forth its strength, and the fruits thereof will be meat for man. The more purity that exists the less strife; the more kind we are to our animals, the more will peace increase and savage nature of brute creation will vanish away. (Kay and Brown, 256)

Early Mormon writings presented a group which, in at least a portion of their rhetoric and land policies, embraced some land conservation concepts. In 1847, Mormons appeared to be poised as enlightened caretakers of a divine gift. They clearly intended to use the Earth to generate monetary gain. However, while capital accumulation was seen as a defense against Gentile intrusions, the Mormons also seemed to develop a unique, theologically based conservation ethic. Brigham Young urged people to behave righteously to appease God and thereby assure plentiful crops, good weather, fertile soils, and healthy livestock. Drought and insect infestations were seen as trials. In Mormon landscape ecology "living right" was crucial in order to earn both God's favor and worldly gain.

Utah was Mormonism's last stand—the end of the exodus. However, the urge to build Zion was a multifaceted force. Mormons saw Zion as three fundamentally disparate geographic realms. They held a vision of Mount Zion and spoke of home in the everlasting hills. Yet the Mormons actually settled in the valleys, where they strove both to construct a city for God and live according to an Edenic agrarian ideal. However, as Easterners, they had no real grasp of the low carrying and regenerative capacities of the sensitive Great Basin and Wasatch Range ecosystems for rural and urban uses. This inexperience and a rapid rise in the demand for resources over the coming decades were to have severe environmental and social consequences.

THE WASATCH MOUNTAINS: MOUNT ZION?

While valley lands were distributed in a fairly democratic fashion, the mountains were managed far differently. Away from the watchful eyes of neighbors in carefully platted cities, the communal use of forests and mountain grazing lands often broke down into a chaos of competition. The LDS Church decided to establish a system of seemingly monopolistic concessions to individuals who would build roads and bridges, cut timber, and otherwise develop the watersheds of the Wasatch Range. These so-called "canyon grants" were, absent an officially recognized government, no more than squatting. However, the same was true for all land "ownership" in Utah, until the opening of the General Land Office in Salt Lake City in 1869, which began to administer land claims under the Preemption and Homestead acts.

Canyon grants were bestowed and administered first by the LDS Church (1847–1849), then by the Provisional Government of the State of Deseret, an attempted Mormon nation (1849–1852), and finally by the Salt Lake County Court (1852–1869). Terms varied. Some grants gave exclusive control of all resources, while others allowed concessionaires to charge tolls to those who entered to collect wood. Local judges had authority to issue sawmill and water power rights and to control timber harvest. The Church and the supposedly separate government agencies believed they could both provide needed wood, range, and water *and* control resource use. They could not.

In 1850, Brigham Young, then both Prophet of the Church and Governor of the State of Deseret, received a canyon grant "in perpetuity" for City Creek. This remarkable grant gave him "exclusive control over timber, rocks, minerals, and water" for $500. His monopoly over a main drainage next to Salt Lake City caused local grumbling and bitterness. Apparently, resources didn't belong to *everyone* after all. Young was firmly in command. He said:

> *Now this is an order for the judges to take notice of; it does not come from the Governor but from the President of the Church [he wore both hats]; put these canyons into the hands of individuals who will make good roads into them and let them take toll from the inhabitants that go there for wood, timber, and poles. (Fox, 116)*

The judges generally obeyed.

Utah: An Elusive Zion

The *Deseret Weekly* reported that Big Cottonwood Canyon had its "upper slopes densely clothed in forest much thick with valuable timber" (July 30, 1856). In the early 1850s, the Young family and others within the LDS hierarchy controlled a great deal of the Wasatch. George Cannon, a prominent Mormon and later a delegate to Congress, recalled:

> *At no time . . . [was] ownership or title bestowed upon any person who might occupy the land. But our canyon roads had to be made and it required some action on the part of the Legislature to induce men to build costly roads into our mountains, and to build bridges over our canyon streams. . . . Grants of this kind were given in the early days of this territory for such purposes, and also for herd grounds and other purposes that local rights might be preserved. . . . We lived in Utah twenty years before the Land laws were extended over us; we had to do the best we could. As soon as these laws were extended over our Territory, we then obtained title to our lands. (Fox, 114)*

However, some local people saw the grants as an unfair use of power and influence.

Timber was cut from the Wasatch Range at prodigious rates despite attempts by the Salt Lake City courts to use sawmill permits as a regulatory device. In 1853, there were some fifty sawmills operating along the Wasatch Front. Coniferous trees were the hardest hit, yet aspen was cut extensively for firewood and roofing shingles. Even Gambel oak and maple were stripped for fuel. Prior to completion of the transcontinental railroad in 1869, the burgeoning Mormon settlements were isolated, and as a result, the ecosystems of the Wasatch Range were intensively exploited. Deforestation was so widespread that by 1870, only Big Cottonwood Canyon had substantial timber reserves left. Consequently, by 1872, Salt Lake City had become highly dependent on coal for heating, and nearly all lumber had to be imported by rail. Descriptions of the city referred to it as an "adobe village."

Overgrazing also began to take a severe toll on the ecosystems of the Wasatch Range. In the early years of settlement, sheep, beef cattle, and dairy cattle all grazed in the mountains in small numbers. Yet by the 1850s, Utes began to voice concerns over the lack of forage for wildlife. In the 1860s, both mountain and valley grasslands were deteriorating rapidly. Mormon Apostle Orson Hyde said, "The range is becoming more and more destitute of grass: the grass is not only eaten up by the great amount of stock that feed on it but they tramp it out by the very roots" (Kay and Brown, 262).

Colonization and the Mormon Landscape

By 1889, over 1 million sheep and 350,000 cattle roamed the Salt Lake Valley and Wasatch Range. Ten years later, while the number of cattle had stayed the same, sheep flocks had grown to over 3 million head. Supplemental winter feeding had allowed huge numbers of sheep to survive. During spring, large flocks moved upslope, trampled wet mountainsides, and grazed perennial grasses at their most vulnerable stage of growth. In summer, overgrazing became so pronounced that bands of sheep on the mountainsides could be spotted from the town by the dust clouds they raised.

Huge flocks of sheep were also trailed through canyons to summer at higher elevations. From the 1850s until the early 1960s, Emigration Canyon was used as a sheep driveway leading to nearby Parleys Canyon. Over 400,000 sheep were grazed in these small canyons each season. The bunchgrass and oak brush range became devastated. One nineteenth-century sheepman recalled:

> *There was no leaf or sprig of any kind as high as the sheep could reach, and the ground was absolutely bare. They ate everything that was green. The greatest damage was done by sheepmen using the top of mountains as a trail in driving sheep—the whole mountain top became a dust bed. (Cottam and Evans, 145)*

At the turn of the century, the Federal Division of Forestry expressed concern over the degradation of the Wasatch environment. In 1902, Albert Potter, a Westerner and range specialist, was sent to analyze the mountains on behalf of the government. He found Big Cottonwood Canyon so denuded "it would be difficult to find a seedling big enough to make a club to kill a snake" (in C. Peterson, 245). He saw 150,000 sheep grazing in one *side drainage* of Big Cottonwood Creek. So severe was the overgrazing and overstocking in many portions of the Wasatch that Potter wrote wryly that "the sheep seemed to live on fresh air and mountain scenery" (in C. Peterson, 245). Potter saw complete deforestation, brushlands grazed to dust, ravines used as log shoots, eroded roads, mine tailing piles, slash and sawdust heaps, indiscriminate and recurring human-caused fires—in short, a scene of environmental collapse.

Deforestation, clearing of oak brush, frequent fires, and grazing radically upset the hydrological balance of many watersheds. Flooding became a problem in the 1870s and increased in frequency and severity. By the turn of the century, central Utah had experienced nine major floods. By 1930, thirteen small Mormon communities south of Provo which had been settled in the 1850s had been abandoned because of recurrent flooding.

The United States's national forests were established in Utah in 1906. Public pressure was on Washington bureaucrats to better manage the West's mountain lands. In Utah, despite much public opposition, the Church supported the formation of national forests. Those in power would later use grazing, timber, and special use permits to exclude business competitors, and this may be part of the explanation for Church support of "Federal" control.

By 1910, eleven national forests had been defined, including the Wasatch and Cache. George Perkins Marsh, in his 1864 book *Man and Nature*, wrote of the need for public ownership of mountain ranges to protect the watersheds on which the region depends. Despite the Wasatch Range's national forest status, the abuses did not abate. Overgrazing continued, and the flooding seemed only to get worse.

In many areas, the 1920s were the peak of sheep overgrazing and habitat destruction. Unrestricted hunting also had severe impacts on wildlife. Grizzly bear, wolf, and many other creatures were extirpated from the state. The Wasatch Range elk herds were wiped out. Deer populations were reduced to a few hundred head.

Conservationists and many hunters rallied in response. Elk from Yellowstone National Park were reintroduced in 1925, and both deer and elk populations boomed in the 1940s and 1950s. Increasing development of valley and foothill areas forced big game species to rely on upslope oak brush communities for browse, cover, and winter range. Today, some 60 percent of this winter range is privately owned.

Despite floods, loss of life, wildlife calamities, mudslides, depleted timber, summer irrigation water declines, sedimentation of irrigation ditches and reservoirs, and visually obvious slope denudation, exploitation of the Wasatch Range did not ease in any significant way until the 1960s and 1970s. Some say this destruction has never really ceased.

The complexity and subtlety of Wasatch Range ecosystems were beyond the ability of Mormon settlers to appreciate. They were humid-biome valley people with no experience in arid lands and mountain environments. However, the rapidity and aggressiveness with which early Utahans devastated the mountains are testimony to both the rigors of settlement and a casting aside of certain portions of their utopian religious philosophy. It was a collision of dreams and reality. Mormons began to cast off stewardship ideals like pioneers on the Oregon Trail forced to dump prized pieces of civilized furniture into the sage and dust.

Stewardship notions, especially in the mountains, were overwhelmed by more powerful cultural signals. Mormons had come to believe that all growth and all exploitation were the same as progress. Ethnocentrism gave them a clear sense that God approved of and guided their actions. There was little reason to make careful use of mountain lands when God would soon return and perfect the Earth as their paradisiacal home. It also appears that Mormons were as susceptible to the allures of profit as Gentiles.

However, something more was involved. Despite the concept of Mount Zion and the scriptural references to the sacred mountaintops, Mormons viewed themselves as a *valley* people. This led to a perceptual and behavioral partitioning of the natural environment. If Zion was earthbound, it would be, at least in the early years, a pastoral lowland heaven. The valleys were, therefore, places of group effort and permanence expressed in honest attempts at sustainable agriculture, cooperative economic ventures, and scores of "eternal" villages. The mountains were Babylon. The Wasatch and other ranges were marked by ramshackle logging and mining camps and wholesale environmental destruction made possible by Church-sanctioned monopolistic canyon grants. It was this shared mental map which allowed destructive exploitation of the mountains to occur. As colonization proceeded, the Wasatch Front, the interior back-valleys, and the southern deserts were the stage where the Saints defined themselves and in doing so created one of the most distinctive landscapes in America.

COLONIZATION: THE SPREADING OF ZION

Brigham Young had an abundance of land-use planning ideas. The Great Colonizer repeatedly rose to the pulpit to call pioneers to leave Salt Lake City and settle new lands, mostly south along the Wasatch Front. Young was the architect of an ecclesiastical Homestead Act. In 1847, only Salt Lake City and Bountiful were settled. However, Young envisioned a State of Deseret, literally an autonomous country, which would cover all of Utah and vast portions of Nevada, Oregon, Idaho, Wyoming, Colorado, New Mexico, and Arizona, and extend to the Pacific Ocean at San Diego (Map 8).

"Deseret" means "Honeybee" in the "Reformed Egyptian Script" of the gold plates of the Book of Mormon. After 1847 the Saints were indeed geographically industrious, initiating many settlements south from Salt Lake

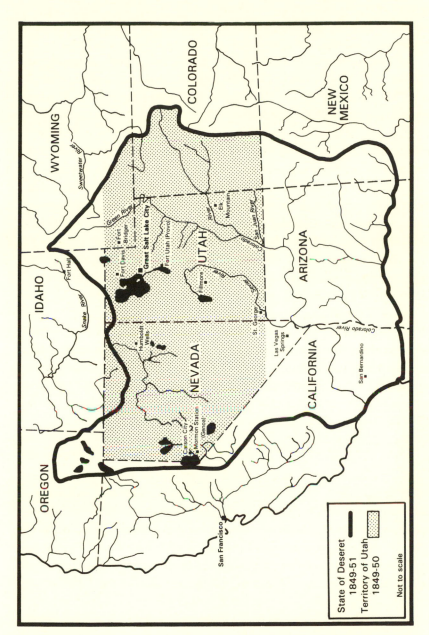

Map 8. Western States, 1847–1869 (redrawn from Campbell, p. 54)

Colonization and the Mormon Landscape

along the "Mormon Corridor." These colonies included Provo, Cedar City, St. George, Las Vegas and Carson Valley, Nevada, and San Bernardino, California. During the "Utah War" of 1857–1858, Federal troops reined in the Mormon quest for an independent Deseret. Brigham recalled the faithful from California and Nevada to the core of the colonies along the oasis at the western base of the Wasatch Mountains.

President Brigham Young relied on a constant stream of new arrivals to provide labor and money for maintenance of the colonization effort. The Perpetual Emigration Fund was established from tithes and was used to pay for the passage of Easterners and foreigners (mostly British and Scandinavians) to Utah. The New Jerusalem was a powerful magnet for converts. Some even resorted to walking to Salt Lake City, dragging their possessions behind them in handcarts.

The population of Utah soared. In 1848, only 1,700 Saints resided there, all in the immediate vicinity of Salt Lake City. Two years later, the Territory had 11,380 non-Indian residents. By 1860, the population reached 40,273, and by 1870, it had more than doubled to 86,786.

Brigham Young died of appendicitis in 1877 and was, with an estate valued at about $2.5 million, the richest man in Utah. Much of his wealth came from real estate holdings, which he largely deeded to himself as a privilege of office. His death did not interrupt Mormon expansion. In 1880, the Territory contained 143,903 people, of which over half were foreign-born. Brigham's vision carried the Mormons forward long past his death.

The new arrivals fanned out into many of Young's Utah colonies and began to spread into adjacent lands. Settlements arose in Arizona along the Salt River (Mesa, Tempe, Lehi), Gila River (St. Johns, Holbrook), and San Pedro River (St. David). Mexican colonies in Chihuahua arose in the 1890s after Prophet Wilford Woodruff's "Great Accommodation," in which he announced that the Mormons were abandoning polygamy. This reversal of policy was done in exchange for Utah's statehood, which was finally granted in 1896. In Mexico, polygamy was not prosecuted, and many church leaders, including Woodruff, maintained "plural" households in this Chihuahuan "refuge."

In New Mexico, Mormons made little headway amid the maze of Indian and Hispanic land grants. The Saints got only a foothold in the northwest corner of the state at Farmington and Ramah. The Colorado settlements were restricted to the San Luis Valley. In Wyoming, Mormon settlements arose around Evanston, in the southwest corner of the state, and in the Bighorn

Basin east of Cody. Most of southern Idaho along the Snake River Plain became unmistakably Mormon. Ely and a few other Nevada communities fell into the LDS sphere. Cantankerous Montana was largely passed over in favor of southern Alberta, which acted as the "Canadian Refuge." An LDS temple was eventually built near Lethbridge at Cardston.

By 1890, national economic depression and increased immigration were severely straining Utah's economy. In order to stem the tide of converts coming to Salt Lake City, church leaders announced that Zion was not just Salt Lake or Utah, but wherever Saints lived righteously and a temple was built. The expansive colonization effort of the Church in the Western United States between 1850 and 1890, often described only as a "Gathering," was in fact also a Mormon diaspora.

THE MORMON LANDSCAPE

Despite widespread colonization, Salt Lake City and the Wasatch Front were and are the core of the Mormon world. They were the cradle of the Church and of what has come to be known as the "Mormon Landscape." This cultural terrain is an agricultural place, mostly valley bottoms and benchlands, containing four-square platted towns with wide streets flanked by ever-present irrigation ditches. Wallace Stegner wrote a defining book about this landscape, which he called *Mormon Country*.

Mormon Country is a land where unpainted barns and granaries stand in town, and the surrounding open fields are dotted with hay derricks and beehives set out in clover and alfalfa. All is bordered by an eccentric, vernacular hodgepodge of sticks, boards, and posts known as the "Mormon fence." Houses are plain and often built of stone, brick, or adobe. Many have small front porches dominated by two chairs. The design of the LDS "Ward House" in the center of town follows a plan issued by the Church Architectural Office. Each period had its official ward style. The town plat always numbers streets North, South, East, and West of the ward, stake, or temple.

The forever upward-reaching lombardy poplars lining the banks of irrigation canals provide a compelling signature of the Mormon way. Compelling, because above all, it was the control of water which made the settlement of Utah possible. Irrigation required of the Mormons something they already practiced and preached in certain aspects of their lives—cooperative effort. The group ethic necessary to build and maintain irrigation systems reinforced Brigham Young's goals of cohesiveness, common labor, and communal enterprise. What arose was something of a theological hydraulic society. A water master, often an LDS bishop, watched over the canals, ditches, and headgates to assure equitable water delivery and to assure that maintenance duties were completed by work crews.

Florence Merriam, a New Yorker who visited Utah in 1894, observed the clustered nature of Mormon settlement and recognized it as an instrument of social control arising from "Brigham Young's shrewd policy of centralization" (Merriam, 3). The Mormon town was truly a social, economic, educational, and religious unit surrounded by a pastoral world. It was a tangible artifact of the Edenic ideal. Even lombardy poplars, so indicative of Mormon agriculture, reveal much about *where* early colonists felt Zion would be built. These tall poplars give a quality to the landscape so distinctive it is still possible to recognize a Mormon town from a distance simply from its biogeography. In *Mormon Country*, Wallace Stegner wrote of these so-called "Mormon trees":

> *They do not grow in the mountains, but neither do Mormons, except for scattered sheepherders and cowpunchers who by their very profession are cut off from the typical Mormon way of life. Mormons and Mormon trees are both valley races. (22)*

For decades, the agrarian Mormon landscape was one of the West's most obvious realms. This was not accidental. The LDS Church believed that authority is best maintained by repetition and redundancy. The spatial form of the cultural landscape itself would remind Mormons *how* to live. This reduced ambiguity and certified that the geographic thought of the Prophet was true. However, such an artifact of spatial Americana would not persist for long in the face of unrelenting population growth, water development, and industrialization. This would be particularly true at the center of the Mormon experience—the valleys along the western edge of the mountains, known as the Wasatch Front.

15

THE WASATCH FRONT:
GEOGRAPHY AND DEVELOPMENT

Brigham Young's 1850 canyon grant for City Creek in the towering Wasatch mountains made him a powerful water broker. Salt Lake City depended on this small creek for sustenance, yet Brigham wasn't about to stop there. In 1856, the Prophet had a foreshadowing vision of the Wasatch Front's future urbanized hydrological landscape (Map 9):

> *Shall we stop making canals when the one now in progress is finished? No, for as soon as that is completed from Big Cottonwood Creek to this city, we expect to make a canal on the west side of the Jordan River, take its water along the base of the west mountains [Oquirrhs], as there is more farming land on the west side of that river than on the east. We shall continue our exertions until the Provo River runs to this city. We intend to bring it around the Point of the Mountain to this city. Even then we know not the end of our public labors and enterprises in this Territory, and we design performing them as fast as we can.* (*Journal of Discourses*, LDS Church, 3:141)

Young imagined diversions on the Weber, Bear, and other rivers which would carry water to an eternally expanding string of cities. The pragmatic leader understood that the most direct way to maintain power in Utah was to control the water. The 121st Psalm reads: "I will lift up mine eyes unto the

∧∧∧∧∧

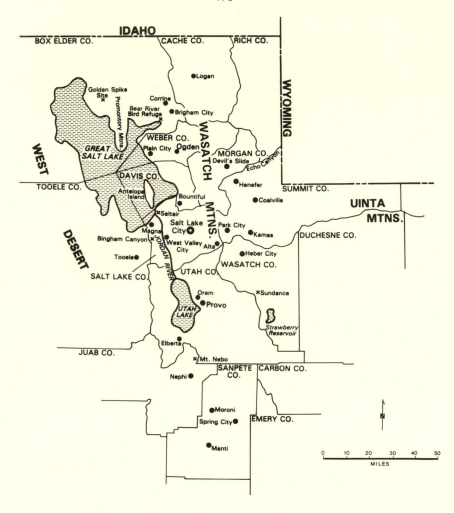

Map 9. The Wasatch Front (redrawn from U.S. Department of Interior, p. iv)

Utah: An Elusive Zion

hills from whence cometh my help." To Mormons of the oasis culture of the Wasatch Front, this help was as much hydrology as heaven.

In 1850, the population of Box Elder, Weber, Davis, Salt Lake, and Utah counties was 11,300, of which 55 percent resided in Salt Lake County. At that time these counties, collectively called the Wasatch Front, contained 93 percent of Utah's total population. Irrigation, livestock, industrial, and domestic withdrawals were underway from numerous mountain creeks and from the Jordan River on the Salt Lake Valley's west side.

By 1872, schemes were proposed to build a diversion dam on the Jordan River and to irrigate its adjacent benchlands. In 1883, the District Court granted Salt Lake City and five canal companies each a one-sixth interest in the water of the Jordan River. At the turn of the century, the population of the 5-county Wasatch Front region had grown to 153,000, half of which lived in Salt Lake County. However, because of the masterful technical and social engineering works of Brigham Young, the state's people had become much more widely dispersed by colonization. By 1900, only 52 percent of Utahans resided along the Front.

The Newlands Act of 1902 authorized formation of the Reclamation Service, now called the Bureau of Reclamation. The agency was charged with moving water hundreds of miles across mountains and deserts to wherever politicians and corporations demanded.

In 1903, the Utah State Engineer, A. F. Doremus, produced a map showing a "proposed intercepting channel" to run perpendicular to the rivers and creeks draining the south slope of the rugged Uinta Mountains. Doremus fleshed out Brigham Young's vision and proposed a system of dams, reservoirs, canals, and tunnels which could ship Green River water to Salt Lake City.

In 1906, the Reclamation Service released a list of its initial projects. Among them was the Strawberry River Project in the Uinta Basin lying southeast of Salt Lake City. A year earlier, the Federal government had opened the high, windswept area up for homesteading. With the promise of irrigation water from a reservoir on the Strawberry, land speculators went wild. The population of Uintah County grew from 6,500 in 1906 to 12,000 in 1907. Wasatch County expanded so greatly that its eastern portion was split off to form Duchesne County. Settlers began to organize new towns as boosterism, based on expectations of plentiful water, spawned an irrational land rush.

The settlers' optimism began to fade when the growing season proved brief and the delivery dates for water were postponed, drought tightened, and debt deepened. However, the Enlarged Homestead Act of 1909 led to continued real estate chicanery in the region, sometimes at the expense of the Indian land claims. The Strawberry River Dam was completed in 1922, and a large reservoir backed up westward into the mountains toward Salt Lake. The initial rationale for the impoundment was local irrigation, some of which actually took place. But as the years went by the specter of inter-basin transfers cast a lengthening shadow across the entire Uinta Basin.

By 1930, the 5-county Wasatch Front region contained 319,000 people, or approximately 60 percent of the state's total. With the population of the Front more than doubling in only three decades, state agencies began to take notice. In 1935, the Utah State Planning Board's *First Report on State Policies* contained farsighted plans for protecting the state's agricultural land from suburban development. The Board advised the state not to engage in "over-promotion of extractive industries" and urged that "all possible means be used to bring back a revival of agricultural pursuits" (155). However, in typical comprehensive planning doublespeak, the Board also stated that the goal of the document was "to put into effect a definite plan or program for the ultimate development and utilization of the natural resources of the state" (228) and offered many ideas for promoting urbanization.

During the 1930s, the Utah State Planning Board feared that the reservoirs and rivers of the Wasatch and other ranges would go dry. These attitudes were deeply shaped by the decade's severe droughts and unrelenting economic depression. The Board recommended that "all possible sources of additional water supply, especially around Salt Lake City and Ogden, be investigated and a complete plan for extension prepared" (47).

Most of the State Planning Board's report contained nuts-and-bolts technical criteria for managing growth and development. However, within the actual *State Plan for Utah* in 1935, specific goals were set for protecting big game winter range in certain areas and zoning was proposed to regulate grazing and to keep erosive lands from being plowed. The Board's concepts for the protection of riparian corridors were extraordinarily enlightened:

> *A serious condition exists in regard to fishing waters, particularly streams because the lands bordering the streams of the state are privately owned. A gradual re-purchase of all stream beds together with narrow strips of adjoining land should be considered. (166)*

However, a system of potentially destructive "Scenic Hi-Ways" was also proposed. One of these roads would have run east-west near the crest of the Uinta Mountains in an area now designated as Federal wilderness. Most of the others have been built, including a road from Hanksville to Natural Bridges (the Hite crossing of the Colorado) and a "skyline drive" along the summit of the Wasatch Plateau. Despite the Board's sincere efforts, little land-use planning actually occurred.

During the 1940s, defense industry jobs spurred more growth throughout the Wasatch Front. The huge Geneva mill of U.S. Steel in Provo, like the facilities at Pueblo, Colorado, was to serve as an easily defended inland source of steel for the war effort. As a spinoff, a massive manufacturing plant was built by Remington Arms in Salt Lake. In 1940, there were only 800 civilians employed on military bases and at defense plants. By 1943, this number had risen to 52,000.

Growth and diversification of the military-industrial complex continued after World War II. Between 1950 and 1963, 30 percent of all jobs created along the Wasatch Front were directly related to defense spending. These industries demanded huge amounts of water. The Deer Creek Reservoir had been created in 1941 as part of the Provo River Project, which was designed to even out annual water supply for both industrial and urban uses. From Provo, water was sent via the Salt Lake Aqueduct northward around the Point of the Mountain and along the base of the Wasatch Mountains to the capital city. The Provo Reservoir Canal and Alpine Aqueduct moved water above the Orem Bench (then agricultural land) north to Pleasant Grove. The last major conduit built was the Jordan Aqueduct, which transferred Provo River water to the Salt Lake Valley along the eastern base of the Oquirrh Mountains through the towns of Riverton and South Jordan.

Passage of the Colorado River Storage Act in 1956 gave Utah's intricate water development projects additional political and financial support. Bureau of Reclamation engineers rubbed their hands in anticipation, since most of the basic plumbing was already in place for the grandest of all their hydraulic dreams: the Central Utah Project (CUP).

In 1960, the Wasatch Front population stood at 691,000—more than twice the 1930 figure. The Front contained 77 percent of Utah's residents, and its demographic dominance continued to rise. Hill Air Force Base had become a major missile center, with companies such as Thiokol Chemical (now Morton-Thiokol), Sperry Rand, Hewlett-Packard, and Browning Ar-

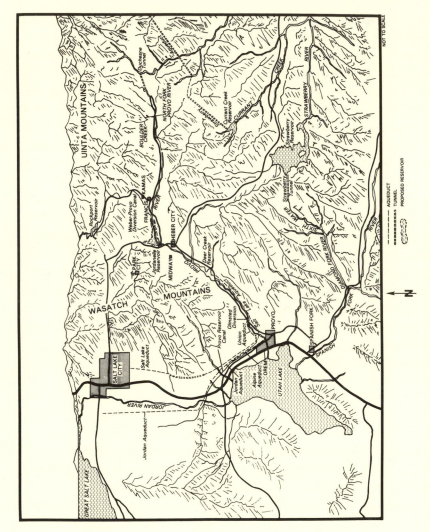

Map 10. Central Utah Project (Source: Central Utah Project)

Utah: An Elusive Zion

maments acting in support roles. The Tooele Army Depot and the Dugway Proving Grounds became major facilities in what is locally called "The West Desert." The demand for water was unprecedented, and the Central Utah Project hit the drawing boards in earnest.

The basic plan of the Bonneville Unit of the Central Utah Project was to move water from Strawberry Reservoir over the hydrological divide into the Spanish Fork River (Map 10). This water would then flow into Utah Lake where it would be stored until it was needed by the downstream cities and canal companies of the Salt Lake Valley. A 1963 U.S. Forest Service environmental analysis found no fault with this scenario. By 1967, the Bonneville Unit officially began with expansion of the Strawberry Reservoir.

During the 1970s, an astonishing population explosion occurred in Utah. The population of the 5 Wasatch Front counties boomed, growing 36 percent from 850,000 to 1,155,000 people. While Weber (Ogden) and Box Elder (Brigham City) counties posted modest gains, Salt Lake County grew 34 percent from 459,000 to 616,000; Davis County (Bountiful) increased 48 percent; and Utah County (Provo) rose by *58 percent* from 137,000 to 217,000. By 1980, the Front was home to 80 percent of Utah's people. Some 85 percent of all state residents lived in urban centers. The thirst of the cities had grown unquenchable, and the need for rapid completion of the CUP was portrayed by politicians as a divine mandate.

THE LAND DEVELOPMENT BOOM

The pace of urban transformation sent many jurisdictions into a tailspin trying to keep up with land subdivision and housing development. In response, the state legislature passed a subdivision law in the mid-1970s which enabled cities and counties to conduct reviews of proposed residential developments. County commissioners were given wide latitude in master planning and zoning: "The boards are authorized and empowered to provide for the physical development of the unincorporated territory within the county." The intention of the lawmakers was to assure that during the building of Zion, streets would be paved and other basic infrastructure attended to. However, the definition of "subdivision" guaranteed that most land splitting would not be reviewed by a planning board.

The Utah Code Annotated refers to subdivision regulation as a way to assure "the protection of both urban and non-urban development," with no mention of conserving wildlife, open space, watersheds, or agricultural lands.

Section 17-27-27 of the Utah Code Annotated defines a subdivision as follows:

> *"Subdivision" means the division of a tract or lot or parcel of land into three or more lots . . . for the purpose, whether immediate or future, of sale or of building development; provided, that this definition shall* not include a bona fide division or partition of agricultural land for agricultural purposes or of commercial, manufacturing, or industrial land [for these purposes]. [*emphasis added*]

Under Utah's subdivision law, exemptions outnumber cases where public review occurs. The exemption for legitimate agricultural subdivision is sensible, but elimination of public input into proposed commercial, manufacturing, or industrial subdivision developments is an astonishing concession to business. There is some review during the enforcement of zoning codes and issuance of building permits, but the public is left out of any substantive analysis of the merits and impacts of potentially dangerous development projects.

For residential subdivision, the review process is triggered by the third division of a tract of record after the date of passage of the county's regulations. Even so, planning boards in most jurisdictions only analyze splits to see that sufficient road frontage is present and that utilities and septic/water improvements are present or can be built. The Utah subdivision law contains no statement of concern for the natural environment, no provision for dedicated parks within developments, and no leadership for the state's people. It is a neutered compromise measure. Allowance is even made for the sale of land by archaic metes and bounds descriptions with no survey plat. However, the law does give counties such wide discretion in adopting and applying regulations that concerned jurisdictions can practice enlightened planning, if they choose.

THE CUP IS CHALLENGED

Utah entered the late '70s in a hail of subdivisions and a prolonged drought. Utah wanted its CUP filled at once. In 1977, President Carter drew up a "hit list" of environmentally damaging pork barrel water development projects. When the Central Utah Project was prominently mentioned as an archetypal Federal boondoggle, the state's elected officials yelped like scalded dogs.

Utah Governor Scott Matheson quickly emerged as an ardent defender of the CUP. In his book *Out of Balance*, Matheson wrote:

> *We are a water poor state, on the verge of massive development and population growth. We are at a juncture which will require our people to be wise with water, to conserve, to alter their way of life in a way that will allow them to survive and compete with the limited amount we have.* (150)

The Governor made no mention of two key alterations of the Utah way of life which would lessen water consumption—conservation and population control. As years passed, the true nature of the CUP gradually emerged. In *Out of Balance*, Matheson explained:

> *The CUP was primarily conceived to provide irrigation water, with some municipal and industrial uses as a secondary priority. But now it is clear that the municipal and industrial uses are making repayment of the CUP possible. It is truly a case of the tail wagging the dog.* (163)

Matheson's apparent naivete is suspect. In 1979, the Bureau of Reclamation published a report showing that 99,000 acre-feet of Bonneville Unit water went to municipal and industrial uses, and only 9,000 acre-feet were set aside for irrigation. Given these widely published data, the meteoric rise of the population of the Wasatch Front, the rapid urbanization of the area at the expense of agricultural land, and the reduction of agriculture to only 2 percent of Utah's Gross State Product, it seems unlikely that as intelligent a person as Governor Matheson could have thought the CUP was anything *but* a scheme to deliver greater supplies of water to municipal and industrial users.

In the late 1970s, Governor Matheson and the state's congressional delegation employed all the influence they could muster to get the CUP off Carter's hit list. Matheson supported Carter's Rube Goldbergesque racetrack plan for deploying mobile MX missiles in exchange for reinstatement of CUP funding. Once the deal was struck, the president became a born-again supporter of the $380 million water project.

Carter was then ambushed as the LDS Church voiced its opposition to the MX deployment. Heavily Mormon Utah and Nevada were both prime sites for the weapon system. The Church's unexpected stand proved fatal.

The Wasatch Front

This opposition, combined with general public outrage, effectively scuttled the nonsensical MX plan. Nonetheless, the tap had been turned back on in Utah.

THE CUP'S PLUMBING

During the 1980s, the Bureau of Reclamation continued to construct the CUP for the local sponsor, the Central Utah Water Conservancy District. The political and design permutations are intricate but illustrate well the grand social and physical scale of the undertaking.

Approximately 126,000 acre-feet of water from the Bonneville Unit's Strawberry River Collection System were projected to be delivered annually to the Wasatch Front via a network of thirty-seven tunnels, interconnecting pipelines, diversion structures, and re-regulating reservoirs. The U.S. Army Corps of Engineers originally refused to issue a Section 404 permit for a portion of the system due to opposition from the Utah Division of Wildlife Resources, the U.S. Forest Service, the U.S. Fish and Wildlife Service, the EPA, and other organizations. These agencies believed construction of CUP would seriously impact many miles of trout fisheries.

During years of lawsuits and negotiations, a compromise was reached which provided for 44,000 acre-feet of water to maintain instream flows on the Provo River and other streams. Off-site mitigation funds were also made available by the Bureau of Reclamation to secure habitats to replace those lost to reservoir inundations, such as the new Jordanelle Reservoir near Heber City in Wasatch County. Because of the compromise, the Army Corps issued its Section 404 permits, and the CUP resumed.

The tentacles of the Central Utah Project continue to reach farther into the landscape. In addition to the main transfusions, there are plans to stabilize and raise small reservoirs in the Uinta Mountains, to build new reservoirs, and to construct hydroelectric facilities. Engineers also proposed to build a dike across Utah Lake to enlarge its water storage capacity in anticipation of the supplemented flows of the Provo River and Spanish Fork. With the reauthorization of the CUP and its hastened construction timetable, water shares in Wasatch Front canal companies have become hot trading commodities.

The original purpose of canal corporations was to secure water for agriculture. With suburbanization and the collapse of agriculture, these organizations are left with little to do. The Central Utah Water Conservancy

District has the power of eminent domain and can condemn water rights which are not being consumptively used. Importantly, it can also claim rights which are deemed essential for urban purposes. Many local canal companies fear that condemnation or "abandonment and forfeiture" proceedings by the Water Conservancy District will "steal" their water shares. Therefore, the incentive for canal interests to sell water to cities is substantial.

Water trading in Utah has grown supremely complex. The water of the Uinta and Wasatch mountains is of much higher quality than that of the muddy Jordan River and is preferred for culinary use. Cities seeking drinking water which hold water shares in the polluted Jordan River/Utah Lake system often strive to sell these shares for rights to the CUP water held in pristine high mountain impoundments. The Central Utah Water Conservancy District is also trying to buy thousands of acre-feet of water bound for polluted Utah Lake and the Jordan River. The Conservancy District plans to siphon the water off at Provo in order to prevent a lessening of its quality and value for residential purposes. Even industrial users are selling water rights to cities. Kennecott Copper (a wholly owned subsidiary of the London-based Rio Tinto Zinc Corporation) holds 108,000 acre-feet of rights in the Salt Lake Valley. Due to more efficient processing techniques at its concentrator and smelter facilities, the company is willing to sell its surplus water to the highest bidder.

In a 1986 letter, Utah Congressman Wayne Owens, usually a dedicated supporter of environmental protection, wrote:

> *Completion of the CUP is, perhaps, the most critical economic development issue facing Salt Lake County and the State of Utah. Without continued firm supplies of water, we will not be able to sustain the growth that is projected for our area and that is needed to bolster our economy.*
> (Salt Lake Tribune, *October 18, 1988*)

The CUP once had a pretense of being a water development effort to benefit an agricultural landscape. This facade has crumbled. In December 1988, Congressman Owens wrote a letter to the Bureau of Reclamation asking that it reexamine the need for the still unbuilt irrigation portion of the system. Opponents of the CUP and pro-growth interests alike criticized the logic of spending $400 million of the $2.2 billion scheme on infrastructure to deliver irrigation water when the remainder of the system was designed to encourage sprawl onto agricultural land.

The Wasatch Front

Farming groups had anticipated this latest redefinition of CUP since 1987, when a bill introduced by the congressional delegation did not include funding for irrigation. Utah's agricultural productivity had always been limited by the state's physical geography. Only 4 percent of Utah is considered arable land. However, by the 1980s, much of this land base had been covered by roads, houses, and shopping malls. Only 2.5 percent of all jobs in Utah are now directly related to agriculture. The state's top employer is the manufacturing industry, number two is government (including Department of Defense jobs), and in third place is the Church of Jesus Christ of Latter-Day Saints. Today, New York has a more agriculture-based economy than Utah.

The communities of the Front are a string of population bombs. In 1988, *The Salt Lake Tribune* ran a front page story entitled "Demographers See Throngs in Utah's Future." The article cited a report by *Population Today* which labeled Utah as "the nation's most extreme example of fertility." By 2050, the report concluded, there will be more people in Utah than in Massachusetts, which currently has a population of over 6 million.

In 1988, the Mormon Church's *Deseret News* published a story titled, "Utah's Fertility Rate is Higher than China's." Fueled by the Church's advocacy of large families, Utah's population growth operates like a breeder reactor. The median age in the U.S. is thirty-two. In Utah, it is twenty-five. Utah has the youngest population of any state in the union and is likely to retain this distinction. U.S. Census Bureau projections place Utah's median age at twenty-seven in 2010, when the national figure will stand at thirty-nine. These mathematics and the mounting demands for water are frightening.

CONCLUSION

Today, predominantly Mormon Utah has clearly demonstrated a theological and cultural preference for large families and the growth of cities rather than crops. The *City* of Zion is the dominant model of modern life in the Beehive State. Over the years, the Central Utah Project had changed purpose, shape, size, and location in response to shifting political pressures and mounting population gains. As Utahans wrestled with their socioeconomic and land-use identity, the CUP was altered and amended accordingly. The 5 counties of the Wasatch Front presently contain some 1,350,000 people, or 80 percent of the state's population. Utah's overwhelmingly urban political power base is clearly in favor of transferring water from the Colorado River watershed or the Arctic if necessary, as long as they can water their lawns, wash their cars, fill their pools, and play at the local water slide park. The Central Utah Water Conservancy District, canal companies, farmers, industries, and cities are playing high-risk water poker in order to cash in on this lucrative, expanding market. It is unknown how long the massive CUP transfusion will meet the Wasatch Front's rising demand. Thus far, there have been no serious attempts to conserve water or to limit growth. In this singular culture, to raise such issues is political suicide and tantamount to religious apostasy. Yet this story of landscape change varies from city to city along the Front and in the sheltering back-valleys lying to the east. The losses mount, but against substantial odds, there are still pockets where people have not misplaced their care for a bountiful land.

16
SALT LAKE:
CARVING UP
THE EMERALD CITY

The isolated Salt Lake Valley was a superb choice for an expanding colony of land-hungry Millennialists. While Denver had been thrown up during the chaos of a gold rush, Salt Lake City was founded as a centrally planned, orderly, religious enclave where snow-capped peaks sent streams running out across fertile soils.

During the 1860s, the first break from the City of Zion plat occurred when a bench northeast of Temple Square was subdivided, and a neighborhood known as "The Avenues" sprang up. In 1871, *The Salt Lake Herald* surmised that:

> *Salt Lake City is destined to be a great railroad, manufacturing, agricultural, and mining center, in other words, a great city, the most populous and influential between Chicago and San Francisco.*
> *(August 25, 1871)*

In 1869, the transcontinental railroad was completed across northern Utah, and a branch line from Ogden soon opened Salt Lake as a regional trade center. For better or worse the Mormons were now part of the national fold. The Utah Southern Railroad was extended from Salt Lake City to the Provo area in 1873. Vegetables and other agricultural commodities produced

∧∧∧∧∧

near Utah Lake could be easily shipped into the city. Salt Lake's era of cultural and economic isolation was over.

By 1876, Utah Western had extended rail lines to the resort beaches along the south shore of the Great Salt Lake. Crowds of city dwellers began to compete with gulls and brine flies for space next to the hypersaline water. In 1880 a Utah Northern line was completed all the way to the burgeoning mining districts of Montana. The Salt Lake City region was booming and undergoing a dramatic transformation. By 1890, the old pattern of villages separated by agricultural fields was rapidly being replaced by a network of cities such as North, West, and South Jordan, Sugarhouse, Murray, Riverton, Bluffdale, Sandy, and Draper.

Water shortages began. When Salt Lake City began running out of drinking water, officials negotiated a water exchange with farmers. The farmers traded rights to high-quality water from Emigration and Parleys canyons in the Wasatch Range for turbid water from the newly completed Jordan River Canal on the west side of the valley. Such trades would prove habit forming.

At the turn of the century, Salt Lake City was a potent banking, manufacturing, and mining center. With industrialization came increasing numbers of Gentiles who worked in the area's mines and mills. Cultural conflict, the Mormons' albatross, occasionally flared up over escalating land purchases by Gentiles. By 1911, Salt Lake City's still dominantly Mormon population exceeded 100,000, and the city had taken on the look of an industrial hub. Although the Church had historically opposed the mining of "nonuseful" metals, the economy of Salt Lake City was at this point wholly dependent on mining revenues. This became boldly apparent when the silver and gold operations in the Wasatch at Alta and Park City closed down, and extraction of copper ore in Bingham Canyon in the Oquirrh Mountains proved to be the city's economic savior.

In 1918, in the face of a spreading smokestack landscape, *The Deseret News* urged:

> *City officials will be wise to pay close attention to the frequently conflicting demands of business interests and residential rights. We have passed the stage when it would be required that too much sacrifice be exacted in favor of commercialism or utilitarianism. Sentiment and beauty, to say nothing of intrinsic rights and values valiantly earned,*

Salt Lake: Carving Up the Emerald City

should now begin to receive attention. In the meantime, let us not become too mercenary. (May 4, 1918)

This column contained shadows of the Mormon Church's former agrarian preference, but it was also a call for land-use planning efforts to protect a suburban, family-oriented way of life.

LAND-USE PLANNING AND CONSERVATION

Industrialization raised several dilemmas. It brought with it not only an expanded economy but a miasma of air pollution. Nationwide, Salt Lake City was known as the "Pittsburgh of the West." With growth came not only an epidemic of lung disease, but an even greater erosion of the Mormon sense of order. In 1926, the Church's *Deseret News* took a startling pro-planning stance:

Salt Lake City has in the past suffered from a lack of zoning. If the zoning ordinance is ratified by the people there can be no complaint. . . . The value of zoning to the development and growth of cities and towns is generally conceded. (January 8, 1926)

Salt Lake City initiated zoning regulations a year later. It would be decades before many other Rocky Mountain cities followed suit. Today, the local land-use system remains focused on zoning. However, throughout Salt Lake's planning history, little consistent effort has been made to conserve ecologically critical portions of Zion.

One of the most notable exceptions is the City Creek Park Project. In 1857, Brigham Young acquired a canyon grant for the entire drainage. He operated a sawmill and flour mill beside the creek, and the LDS Church Relief Society established a silk production facility using worms imported from China. During the 1860s and 1870s, Young sold or gave away subdivision lots in the lower watershed to family and friends who built many grand homes. In 1902, the Salt Lake City Council passed a resolution proposing that City Creek be set aside as a park and began to acquire land in the watershed. Today, the neighborhood around City Creek is one of Salt Lake's most unique. The large Victorian homes, tall shade trees, and beautiful flower gardens are an island of tranquility immediately adjacent to the downtown skyscrapers. Fifty-two houses have been included in the City Creek Historic District.

The 1986 City Creek Master Plan stresses that the canyon "should serve as a valuable watershed and recreation/open space amenity" (City Creek Master Plan, 1). The plan outlines farsighted policies for preserving the historic district and conserving the mosaic of city, county, U.S.F.S., and private lands in the upper reaches of the watershed. However, problems have arisen with some of its recommendations for land acquisition.

Although the city pledges to "devise a long-range strategy for acquiring all privately-owned property in the canyon," its plan makes no mention of the most widely used land protection tool in America—conservation easements. Land exchanges are given only a cursory nod. Mainstream funding sources such as real estate transfer taxes, general obligation bonds, and an open space fund receive no recognition at all.

The master plan announces that the city will try to "prohibit access to and development of privately-owned property in the canyon." This intention is not only unfair, but illegal in most states. Tactics aimed at blocking access to private land are not only an inequitable way to conserve the canyon, they are political hari kari. Nationwide, negotiations with private landowners and the U.S. Forest Service for conservation and trail easements, land trades, and recreation management agreements have proven far more pragmatic, popular, and appropriate. It is also extremely unlikely that the city will achieve its goals pursuing "the possibility of a Forest Service land grant transferring ownership to Salt Lake City."

However, some progress is being made. In 1986, Salt Lake City, which owns 56 percent of the canyon, classified the upper watershed as a "nature preserve." Given that 15 percent of the city's water supply comes from the drainage, there is also a decidedly practical rationale for protecting this resource.

Despite procedural problems, the City Creek Master Plan is an excellent design for future use of the watershed. A trail system is partially in place and extensions are underway. The creek is a fine place to birdwatch and offers amazing solitude only five minutes from downtown Salt Lake. The canyon is also an extremely popular riding area for thousands of local mountain bikers. City Creek Park will take years to bring to fruition, but it is, even though incomplete, already serving important recreational, open space, historic, and ecological functions.

Salt Lake City planners indicate that access to public land along other portions of the Wasatch is also a major priority, together with limiting urbanization on steep slopes. One planner says, "It's easy to see where conservation is needed in the foothills—there's just not much deer and elk winter range left up there on private land."

Salt Lake City has a Hillside Protection Ordinance which prohibits housing development on areas with 40 percent or greater slope steepness. There is also a policy of trying to keep development below the 5,200-foot contour. These efforts to protect wildlife habitat and open space within the Foothills Preservation Zone are seen most clearly in the Avenues Master Plan.

The Avenues is a district of historic homes lying east of City Creek. Mark Lindsey and his fecund wife, Berthiah, operated a picnic area and bathhouse there during the 1860s and 1870s. In 1872, they subdivided their land into the tracts which today comprise the neighborhood. The city is presently encouraging large-lot residential development and hillside protection in the nearby grass- and oak-mantled upper slopes. The area's master plan refers to "Properties Preserved by Access and Zoning." Access closures and zoning are tenuous and eminently reversible stopgap measures. Certain other lands are classified as "Preserved by Easements." Planners describe these lands as "deeded easements to the city in exchange for development approval on the rest of the parcel." In other words, these "easements" are really agreements with developers which, while effective, are much different than the conservation easements used by land trusts.

The Avenues plan also contains a map of "Properties Recommended for City Acquisition." According to local land-use planners, Salt Lake City is considering disposing of some of its land holdings to "establish an on-going funding source for open space acquisition efforts." This is currently one of the most viable ways for the city to begin conserving key lands in places such as the nearby East Bench neighborhood.

The East Bench is an upper–middle class district of older homes in the familiar four-square Mormon grid. Above Foothills Drive is an enclave of newer homes backed up against the mountains on winding streets and cul-de-sacs with names such as Oak Hills Way, Oakhurst Drive, and Oak Springs Drive. When the Salt Lake area was agrarian, Gambel oak was seen as a scrubby plant of waste areas—the vegetable equivalent of vermin. Today, only the well-to-do can afford to live in the oak brush ecosystems where the views are superb and the land values high. The recurrence of "oak" in street signs in exclusive districts and in the names of high-ticket development projects is not accidental. Gambel oak has become a keynote species of the region's land marketeers.

Utah: An Elusive Zion

The East Bench Master Plan stresses the need to protect the neighbor-hood's geologically unstable, oak-dominated mule deer winter range habi-tats, which also serve as a scenic backdrop for the city. Although the East Bench Master Plan's conservation goals are well conceived, the plan's imple-mentation techniques are shaky:

> *Owners of foothill property should deed all undevelopable property to the city as a condition of approving the subdivision of developable por-tions of their property. Considerable tax incentives exist for such property donations. (East Bench Master Plan, 5)*

When such a donation is a mandatory condition of subdivision approval, the landowner's claim to Federal tax incentives for "charitable gifts" is elimi-nated. Even if this were not the case, a person deeding "undevelopable prop-erty" would hardly receive "considerable tax" savings, since such lands would have little market value.

Salt Lake City is making an effort to protect land, establish access points and a trail system, and create foothill parks in the East Bench area and other neighborhoods. In 1988, the city prepared an Open Space Plan which con-tained an inventory of existing parks and plans for a trail system. The foot-hills zoning and the Hillside Protection ordinances are achieving some suc-cess. Yet, possibly illegal attempts to deny access to private land, imaginary "land grants" from the Forest Service, and inaccurate presentation of conser-vation easements and Federal tax law simply will not further the city's excel-lent goals for the protection of open space.

THE RURAL FRINGE

Salt Lake County is much like a worm-eaten apple. The growth of existing cities and the incorporation of new ones have caused continuous annexation of more and more formerly rural areas of the county. Of the 715,000 people in Salt Lake County, only 280,000 actually reside in areas receiving exclu-sively county services. The infilling pattern is less sprawl than splatter. The annexation hostilities over prime tax-generating industrial and commercial areas resemble the Peloponnesian Wars. About all that is left of Salt Lake County is a scatter of far-flung unincorporated, unannexed lands. Much of this is residential areas which are a net tax liability, since it costs more to provide services to these bedroom communities than they generate in revenues.

Salt Lake: Carving Up the Emerald City

Salt Lake County planners believe that since the planning effort has been split up into nineteen community districts, people have more direct input. The downside of this grassroots level planning is that it lacks an eye toward the whole. While the county has agricultural zoning which limits houses to densities ranging from one house per acre to one per 20 acres, the technique has proven ineffective as a land conservation device due to local resistance to regulating your neighbors. A local planner concludes:

> *Zoning is not really stopping agricultural lands such as wheat fields and orchards from being converted to housing, industrial, and commercial uses. There is too much political pressure against land use regulation to make it stick. As soon as people want to do something different with their land, the process breaks down through petitions for rezoning, variances, or they are annexed into a city, and the zoning is changed to encourage development.*

A drive around Salt Lake County shows how true this is. West Valley City, the third largest city in Utah behind Salt Lake and Provo, didn't exist until the early 1980s. Once a fertile center of truck farming, the area is now a crowded industrial and warehousing smurge. Transshipment facilities of Consolidated Freightways and Transcon are mixed with mobile home lots, industrial parks, car washes, and clusters of fast food stores. A few undeveloped parcels exist north of West Valley City out toward the Salt Lake airport. Large powerlines crisscross above modest suburban tract homes. A few people have horses on their 2-acre ranchettes. The dominant sounds are dogs barking, traffic, and jet wash from the airport.

In a recent editorial in *The Deseret News*, a local couple asked, "Who is Running West Valley City?" Their comments are a litany of the problems associated with instant cities:

> *Residents of West Valley City were under the impression that incorporation would bring government closer to the people. . . . Decisions are being made on land development helter skelter. Commercial development is being approved in residential areas. Not only are we getting a glut of convenience stores we now have a glut of starter homes. Dead carp and other dead animals lie along the roads because of a poorly managed water drainage system. This is ridiculous! (December 4, 1988)*

Utah: An Elusive Zion

To the south, it is impossible to tell West Valley City from Kearns or Bennion. West Jordan stands out only because the huge surrealistic LDS Jordan River Temple is located in a field outside town. On this polluted winter day, the temple's angular white spire and broad variegated base look like Angkor Wat looming in a tropical haze. The pastures in front of the temple remain open. However, two signs foreshadow the demise of this sheltering open space—"Available: 17 Acres for Sale" and "Proposed Site of Cornerstone Condominiums."

This is also largely true in South Jordan despite a weak system of agricultural zoning. Signs placed in the front yard of a farmhouse and orchard exclaim: "NOW OFFERING PEACH BLOSSOM: A Quality Planned Community." Once again, a subdivision destroys its namesake.

THE COPPER LANDSCAPE

The Bingham Canyon open pit copper mine in the nearby Oquirrh Mountains is one of the few human artifacts visible from space. The gigantic flat-topped piles of overburden loom over the west side of the Salt Lake Valley and dwarf the ancient Lake Bonneville shorelines. Ore was originally discovered there by Erastus Bingham, who grazed cattle in the canyon during the 1840s. In the 1860s, *The Deseret News* responded to plans to mine copper:

> *In these days gold is the principal thing sought after, and a man who would engage in copper mining might be considered in a state of insanity. (in Hinton, 171)*

Insanity of a sort, perhaps.

The Bingham Canyon Mining District was formed in 1863, and ore began to be extracted from underground tunnels. On a hot August day in 1906, the Utah Copper Company unleashed steam shovels to begin excavating an open pit mine. Overburden was stripped at the rate of 100,000 tons each month. A gifted metallurgist named Daniel Jackling had conceived a way to commercially extract copper from very low grade porphyry copper ore. As a result, by 1909, the Bingham Community Club boasted the pit was producing "the greatest copper tonnage in the world—Bingham will be active when every other camp has passed into oblivion" (Hinton, 171). Jackling was a state hero, and local miners called him "the King of Porphyries." Utah historian Leonard Arrington declared him "the Henry Ford of copper mining."

Salt Lake: Carving Up the Emerald City

In 1915, the Kennecott Copper Corporation was formed and began to operate the mine. Kennecott bought the ASARCO smelter at nearby Magna in 1959 and began unified copper production under one company for the first time. Long a major polluter of the Salt Lake Valley's airshed, Kennecott spent $280 million in 1978 at the Magna smelter to meet EPA air quality standards. From 1985 to 1987, the facility once again shut down for modernization. Full-time operations have resumed with 200,000 tons of refined copper and substantial amounts of gold, silver, and molybdenum produced yearly. Some 2,200 workers are needed to keep the system up to speed.

Photographs of the Bingham Canyon pit leave you unprepared for the scale of excavation. Kennecott has created a stunning topographic reversal. Where a mountain once stood, there is now a crater over 2-1/2 miles wide and a half-mile deep. If Chicago's Sears Tower were placed in the bottom, it would look like a child's maypole. The meandering strips which encircle the vast pit and continue up the adjacent mountainsides resemble the eroding played-out agricultural terraces of a vanished culture.

The mind has trouble grasping the enormity of it all. A company pamphlet describes it as the largest open pit mine in the world. Ore-hauling trucks, which appeared in the distance as orange specks, later rumble past with ten-foot-tall tires. The man in the cab looks like a mouse driving a Rambler.

The financial return has been equally astonishing. The pit's red metal has generated eight times the economic value of the flashy gold and silver strikes at the Comstock Lode, Mother Lode, and in the Klondike. However, a price has been paid. The landscape for miles around Bingham Canyon has been so transformed that reclamation seems impossible. The devastation is complete.

North of Bingham Canyon on Route 111 is the enormous Bacchus complex. The red and white checkered water towers are reminiscent of the corn and soybean landscape of the Midwest. However, Bacchus is where missiles are assembled for the Department of Defense. The grain fields and pastures around the plant are not protected for their scenic or ecological importance, but serve as safety buffers in case of massive explosions.

In Magna, the scene is equally grim. Sullen brick and metal smelters and concentrator buildings squat at the base of devegetated hillsides. The area is a tangle of railroad tracks, conveyor belt systems, and electrical transmission towers. The Chevron Chemical Company plant, Bonneville Raceway, welding outfits, and heavy equipment lots border the smelter complex. Copper blast abrasives, railroad ballast, and construction aggregates are available at the nearby UP Resources yard. A 4-mile by 3-mile tailings pond lies north of Magna at the site of an old Pony Express station, about 1 mile from the shore of the Great Salt Lake.

A Morton Thiokol salt/fertilizer plant sits distressingly close to the tailings dump. In summer, the pond dries up, and dust from the smelter slag blows toward Magna. Due to these airborne particulates and heavy metals, the EPA has warned local residents to wear masks on windy days like this. No one does. The towering smelter stacks stand above the scene at the north end of the Oquirrh Mountains like the Pillars of Hercules on the edge of the wine dark Great Salt Lake.

Main street in Magna is a sad collection of bars, pool halls, liquor stores, and abandoned storefronts. Colisimos Market seems to be doing well, but Papanikola's Furniture Store and the Gem Theater are boarded up. Like most mining towns in the Rockies, Magna is dour and unrelentingly depressing. It's the end of the line. A faded Chamber of Commerce billboard from a more hopeful era proclaims, "Magna: World's Copper Center." On cue, the driver of a passing Ford pickup heaves an empty beer bottle at the sign, sending a shower of brown glass into the foul air.

THE GREAT SALT LAKE

Since the mid-1800s the beaches along the south shore of the Great Salt Lake have been immensely popular recreation attractions. Today, the Great Salt Lake State Park and the Saltair Resort are the main draws. In 1982 and 1983, the unfortunate combination of a state cloud-seeding program and intense El Niño rains resulted in a record-high lake level. In the spring, water poured out of the Wasatch Range so quickly that sandbags were used to direct the floods through the streets of the capital city into the Great Salt Lake. Scores of kayakers jumped the barricades to take the ride of their lives. Children were seen trying to snag trout from downtown arterials.

The onslaught flooded the onion-domed Saltair building and threatened the Southern Pacific Railroad causeway across the lake, the Western Pacific bed, and Interstate 80. During 1986, the lake rose even higher and submerged the automobile causeway to Antelope Island. The lake inundated not only saltgrass and shadscale communities but upslope fescue pastures as well. In 1987, the U.S. Geological Survey measured the lake at a record height of 4,212 feet, some 20 feet above the 1963 level. Because the lake has a flat bed, this "high stand" increased its area by some 900 square miles.

Salt Lake: Carving Up the Emerald City

A year earlier, Governor Norm Bangerter initiated the West Desert Pumping Project, which was designed to lower the lake. Three huge sixteen-cylinder pumps were installed along the west shore at Lakeside. Water began to be pumped through the Rio Buenaventura Canal westward out into the Newfoundland Evaporation Basin. Critics of the $60 million effort dubbed this basin "Lake Bangerter." Pumping continues to slowly lower the lake. The state's goal is to drop the Great Salt Lake by at least 4 more feet.

The littoral of the Great Salt Lake is not reminiscent of the beaches of Jamaica. The mudflats put out an anaerobic stench. Winter waves fetch up berms of salt froth, dead brine flies, and drift trash. Only the exposed root systems of previously flooded shadscale stand above the saline ooze. A local man offers, "The lake isn't used much, it stinks, the brine flies are a pest, and the water makes your eyes burn like someone threw lye in them, but the birds like it, and it sure is pretty." And yes, you *can* float effortlessly in the ultrasaline water. However, if you swim in the stinging fluid with so much as a mosquito bite on your body, prepare yourself for pain.

Most of the magnificent marshlands and bird refuges are on the lake's east and north shores, where the Ogden and Bear rivers provide fresh water. Near Saltair, thousands of California gulls shriek and whirl overhead like a gathering storm whenever children toss pieces of Wonder Bread or Cheetos into the air. Ever since the gulls saved the Mormon pioneers' crops by devouring millions of grasshoppers, there has been a special tie between the two. The gull is the Utah state bird, and there is a golden statue honoring the species in Temple Square.

On still, grey days, the surface of the Great Salt Lake is like malleable glass stretched into smooth waves by the slightest breeze. The light is often illusory in the high, salty air. Clouds, sky, lake, and mountains blend into impressionistic patterns. The marshlands ringing the southeast shore of the lake are inviting places to visit, yet often prove difficult to reach. The wetlands near the Salt Lake Airport, west of Eddie Rickenbacker Drive, are the exclusive domain of the West Side Duck Club. Much of the lake's saltmarsh is privately owned and carefully guarded. Finding access to the Farmington Bay State Waterfowl Management Area is also difficult. Most roadways dead-end at gates marked "Duck Club Members Only." In many places, discarded couches, mattresses, and household debris lie beside the dirt roads. However, wherever fresh water flows into the lake, there is a profusion of waterbirds and other life. But there are limits. An optimist once tried raising lobsters where the Jordan River enters the lake. All the crustaceans went clawsup in two days. The lake itself supports only brine shrimp, brine flies, and protozoa. It is too harsh to nurture even a Salt Lake Monster. While sightings occurred in the 1800s, there seems to be no persistent local folklore about fantastic aquatic beasts swallowing up bobbing Mormons. Apparently not even myths can withstand the lake's choking salinity.

THE JORDAN RIVER

The Jordan River flows north from Utah Lake near Provo through the west side of the valley before entering the Great Salt Lake. Early accounts of the river describe it as a fresh, trout-filled stream bordered by willow trees and grassy meadows. In Brigham Young's time, it was a cool, shady place for swimming and boating. Young mentioned the Jordan as one of the valley's natural resources which should be preserved for recreational use. However, with accelerating population growth and the severe floods of 1862, 1909, and 1917, use of the river shifted to more utilitarian purposes. Its banks were riprapped, and its channel used as a conduit for irrigation inflows and outflows and as a storm drain and waste canal for adjacent communities.

In 1941, Alex Buchanan of the Salt Lake County Planning and Zoning Commission reminded local people that "the Jordan River could still become an asset to this city if it were taken care of" (*Salt Lake Tribune*, August 3, 1941). Three years later, Roscoe Boden, the chair of the County Commissioners, budgeted $800,000 for the river cleanup work. Despite this action, old cars, barrels of toxic waste, effluent, agricultural biocide runoff, industrial discharges, and trash continued to find their way into what amounted to Utah's Love Canal.

In the late 1960s, the Army Corps of Engineers made a proposal to channelize the Jordan River to control flooding. Concerned with the drastic impacts of the Corps' "final solution," the Salt Lake County Commissioners hired Urban Technology Associates to prepare a study of alternatives. The consultants produced a fine plan which addressed recreational use, floodplain management, storm drainage, and pollution control. The plan's central concept was to create a large riverside park system to transform the Jordan from a liability to a public asset. Practical land acquisition tools were described, and suggestions were made on transportation planning, park design, and revegetating riverbanks. The simple wisdom of the plan succeeded in scuttling the channelization project.

In 1977, the State Department of Parks and Recreation made an initial appropriation for land purchases within the newly created Jordan River State Park. Funding has totaled more than $20 million to date. The park is now about 12 miles long, with significant stretches in public ownership. There are plans to continue acquiring land and working with communities to create a system of parks and trails from the Great Salt Lake to Utah Lake, all the way to Provo's parks, and up Provo Canyon to the state-owned recreational train, the Heber Creeper.

Jordan River State Park hosts a surprising diversity of recreational experiences. Two golf courses, a frisbee golf setup, canoe docks, a bike motocross course, and sports fields, as well as hiking and equestrian trails, are spread out throughout the corridor.

The natural environment is far from restored. At Glendale Park near West Jordan, the river is a turbid and slow-moving ditch. The Raging Waters waterslide complex is built beside it. People no doubt come down to the river in the summer to cool off. Finding it too polluted, they splash around instead in city water within the park concessionaire's pools and slides. North of Glendale Park is what is left of one of the Wasatch Range's nicest streams, Mill Creek. The creek is a completely channelized conduit for green, foaming industrial ooze which flows inexorably into the Jordan River.

The Jordan River State Park is a case of making the best of a bad situation. The water pollution problem is perhaps the most intractable issue the state faces. The successful land acquisition program thus far has not included any conservation easements or land exchanges. The mosaic of state, city, county, and private land within the corridor invites the use of techniques other than outright purchases.

In 1988, local conservationists formed the Salt Lake Regional Trails Council. Its members hope to create a network of trails along old aqueducts and canals, abandoned road rights of way, railroad lines, powerline corridors, and undeveloped private lands. Eventually, they believe, it will be possible to hike or bike from the west valley along the Jordan River to the east benches and into the Wasatch Range. The coordination of the Jordan River State Park efforts with those of the Trails Council may one day result in something akin to a local land trust.

THE WASATCH MOUNTAINS

Since the era of canyon grants, land-use practices in the Wasatch Mountains have been the subject of intense local debate. Once the setting for destructive grazing, logging, and mining activities, the Wasatch is now a nationally renowned downhill skiing center. Resorts and other facilities dominate several watersheds. In recognition of this high degree of development, Salt Lake City was named the United States entry in the competition to host the 1998 Winter Olympic Games. The bid failed.

However, the prospect of a Utah Olympics did revive long-standing land-use planning controversies. The most heated disputes are focused on resort expansion within U.S. Forest Service Special Use Permit areas, land subdivision and housing development, and watershed degradation. A "Ski Interconnect" system has been proposed by the ski industry which would link the Alta, Snowbird, Brighton, and Solitude resorts of Little and Big Cottonwood canyons with the Park City and Deer Valley ski areas in adjacent Summit County. The expansion of these ski areas has a direct bearing on the stimulation of economic growth and land development throughout the entire Wasatch Front. Even Utah's license plates shout "The Greatest Snow on Earth" next to the image of a skier blasting through the Wasatch's famous powder.

In the late 1980s, several long public hearings were held on the proposed Wasatch Canyons Master Plan. The same familiar battlelines were drawn. The plan noted "under the [existing] Forest Service Plan, ski resort expansion within existing resort boundaries could amount to . . . a *46 percent* increase over current canyon ski resort capacities." Planners accepted this growth increase but warned that any "ski area expansion beyond U.S. Forest Service permit boundaries would be inconsistent with this plan."

Despite this more than reasonable allowance, the chairman of the Salt Lake Chamber of Commerce called the plan "strident and prejudiced." Several speakers characterized the plan as too restrictive and went on to rail against all attempts to interfere with complete development of the Wasatch. Kent Hoopinggardner of the Snowbird Resort complained about the plan's supposedly extreme bias against ski areas and concluded that its entire tone was inappropriate for a planning document. Conservationists asked incredulously, "Isn't a future 46 percent expansion of ski capacity enough, given that the resorts aren't even operating at full capacity now?"

The Wasatch Canyons Plan also offers several rational recommendations on controlling housing development. The plan advocates limiting new housing to environmentally capable sites, greater use of clustering via planned unit developments, retaining large-lot zoning, and implementing county ordinances requiring trail easements through subdivisions to public lands.

These guidelines have received a surprising amount of support. At one public hearing, State Senator Francis Farley warned that stronger language was needed to prohibit development which damages water quality, saying that it would be a desecration not to protect the Wasatch. Linda Moore of the Utah Division of Environmental Health echoed this concern by reminding the assembled that most of the population of the Wasatch Front consider water quality as their main concern relating to increasing development of the mountains. A 1988 poll taken by the Salt Lake City Public Utilities Department revealed that 61 percent of the respondents said clean water was the most important benefit of the canyons, 31 percent said recreation, and only 6 percent said economic development.

Most of the speakers at the hearings paid little attention to the actual contents of the plan. The process dissolved into high dramatic comedy. As usual, conservationists and developers were speaking two different languages. This collision of values is best illustrated by the comments of the marketing director for one of the major resorts and a member of the Park City Chamber of Commerce. He asked, "If you were selling a copy machine, wouldn't you grab an opportunity to sell more?"

The Wasatch Canyons Plan recommends that "Salt Lake County should establish a program for acquisition of private property." Recognition that a conservation real estate approach is needed may be the beginning of a reasonable dialogue over the location of future developments and protected lands. However, previous conservation negotiations have proven to be intricate.

Little Cottonwood Canyon is best known as the setting of the Alta and Snowbird ski resorts. In the 1860s and 1870s, Alta was a silver boom town with a population of 8,000. Fire and avalanches destroyed the mining town on several occasions. To take advantage of the annual 300 inches of powdery snow, the first chair lift opened at Alta in 1938. Today, the resort is surrounded by the U.S.F.S. Twin Peaks Wilderness Area. It has eight double lifts, thirty-nine runs, and the capacity to handle 8,500 skiers an hour. The area totals some 500,000 skier visits each year. Nearby Snowbird does an equally booming business. Nippon Airways now offers package trips during the winter months from Tokyo to Salt Lake City. Why ski Sapporo slush when Wasatch powder is so accessible? As a result of national and international attention, the mouth of Little Cottonwood Canyon is being rapidly filled with townhouses and condominiums. Fittingly, a nearby marker praises "F. A. Hofman and Company Real Estate Office," which opened on the site in the 1800s. It is perhaps the only roadside historical marker in America paying tribute to a realtor.

Utah: An Elusive Zion

In 1982, the San Francisco–based national conservation organization, the Trust for Public Land, the U.S. Forest Service, and Salt Lake City worked out a creative way to conserve the famous 545-acre Whitmore Estate near the mouth of Little Cottonwood Canyon. This ecologically rich land served as important watershed and had been used as a popular rock climbing site for decades. When the landowner died, his thirty-three heirs, who couldn't afford to hold the land, initiated discussions with developers interested in the $3.9 million property. The mayor at the time, Ted Wilson, had spent his youth pioneering routes up the granite faces of the estate's canyon walls and grew concerned that the place would be lost. He approached the landowners and obtained their promise to hold off on any subdivision projects until conservation alternatives were fully explored.

The Trust for Public Land (TPL) was called in to prepare an acquisition plan. They quickly negotiated an innovative agreement with the Wasatch National Forest. The plan called for TPL to buy the 545 acres and then trade it to the Forest Service in exchange for an equal value of Federal surplus land elsewhere in the state. It seemed a simple solution. However, the surplus property first had to receive ecological, archaeological, and mineral analyses to assure it was of low public value. After receiving these "disposable" Forest Service tracts, TPL would sell them to interested buyers as reimbursement for the trust's original purchase of the Whitmore Estate. However, due to unforeseen administrative and financial hurdles, it took years for the deal to be completed. Today the 545-acre parcel is in Federal ownership and managed as a reserve. It is a monument to the power of persistence.

CONCLUSION

Since 1985, when Utah passed its Land Conservation Easement Act, The Nature Conservancy has been increasingly active in the state. To date, there has been little or no interchange between TNC and Salt Lake City or County land-use planners. With its mission of conserving biological diversity, TNC has worked mostly in the West Desert, Southern Utah (locally called Utah's Dixie), and the Colorado Plateau. TNC points out that it would be a less than optimum use of its time and money to focus on the urban Salt Lake area and Wasatch Front. There is so much of biological importance elsewhere in the state where there is a better chance at establishing some meaningful preserves.

The Salt Lake City metro area is a place of extremes. The Jordan River parkway, City Creek Park, watershed acquisitions, and recent planning efforts are examples of a willingness to conserve resources. However, state, county, and local governments are not pursuing protection of open space, agricultural land, wildlife habitat, and rare species in any comprehensive or carefully considered way. Understandably, The Nature Conservancy's attention is placed elsewhere. Thus far, there are no local land trusts to pick up the slack. As a result, both uncontrolled development and planned development are carving up the very core of the Mormon world.

17

PROVO:

HAPPY VALLEY HANGS ON

In 1776, while Revolutionary War battles raged back East, Fathers Dominguez and Escalante quietly moved down what is now called the Spanish Fork River and entered an extraordinary basin dense with Ute Indian encampments. Utah Lake provided the Indian people with trout; the marshes were full of waterbirds and eggs; and the benches and foothills were rich with grouse, deer, and elk. The Spanish Fathers passed peacefully through this biological Eden truly impressed with its bounty. Fifty years later, Etienne Provost, a fur trader from Santa Fe, came to this legendary Indian stronghold to trap. Word filtered out of a spectacular new land, and Provost was soon joined by scores of other trappers. In 1849, the first Mormons moved into the valley, and two years later, Provo (named for Provost) was incorporated as the hub of a productive agricultural landscape. The Utes soon were forced out and trudged off to less fruitful lands.

During the 1860s, private canal companies were established to transport irrigation water to eager farmers. At the same time, agricultural interests in more populous Salt Lake County built a dam on the Jordan River and began using the enlarged Utah Lake as a reservoir. The rising lake flooded Utah County farmlands and wildlife habitats. In 1885, after years of conflict, a compromise agreement was reached to prohibit raising the lake level above a set elevation of 4,489 feet. The controversy would not end there.

In 1913, the Strawberry Reservoir was completed in the Uinta Basin east of Provo, and some 40,000 acres in Utah County began to receive irrigation

∧∧∧∧∧

water from this impoundment. The resulting increase in total sales of fruits, vegetables, hay, grain, and livestock placed the county in the top one-hundred agricultural districts in America. Life moved along at a steady, prosperous pace.

The bombing of Pearl Harbor shook this quiet farming valley out of its languor. On December 22, 1941, county commissioners formed the Utah County Planning Board in response to rumors of an impending War Department facility to be built near Provo. The Commissioners feared a defense plant would mean construction of substandard worker housing, the loss of prime agricultural land, and skyrocketing county services budgets. A Uniform Building Code and a zoning ordinance were enacted. The zoning measure required a 1-acre minimum for building sites in rural areas as a way of encouraging development within existing cities. Subdivision regulations mandated that developers build sewer and water systems and roads. Utah County braced itself for change.

In 1942, the War Department purchased 1,600 acres next to Utah Lake west of Orem and began construction of a $200 million steel mill. The site was chosen because of its plentiful water supply, reliable Mormon workforce, and strategically safe interior location. War planners felt that the steel mills of Fontana, California, were vulnerable to Japanese bombing. Plants in Pueblo, Colorado, and in Utah were meant to assure a continuous supply of steel for shipyards and aircraft factories. The impact on the valley was immediate and changed the course of Utah County's economic destiny.

With completion of the plant in 1943, hundreds of local workers were hired, and hundreds more moved to the area. As the service and retail sectors grew, farming began to decline. Rural residential subdivisions and spreading cities covered more and more agricultural lands. Utah Lake became severely polluted by industrial discharges. There had been a time when the steel mill site had been so beautiful, a lakeside resort had been built there. The hotel and beach complex was named Geneva, since the presence of the lake and spectacular glaciated mountains reminded the owner of Switzerland. The valley's air, once scented by peach and apple blossoms, turned brown and acrid from foundry smokestacks. In 1946, the Federal government sold the Geneva steel mill to U.S. Steel (now USX). On August 31, 1987, Geneva Steel (Basic Manufacturing and Technologies Company of Utah, Inc.) bought the plant from USX. The company now employs 1,600 workers at one of the West's largest and most polluting integrated steel mills.

Utah County is today a place pulled by strongly conflicting impulses.

Of 255,000 local residents, only about 15,000 live outside incorporated areas. Yet most people perceive themselves as rural despite living in a string of cities along the base of the Wasatch Range. North of Provo are Orem, Pleasant Grove, American Fork, and Lehi. South of the county seat lie Springville, Spanish Fork, Payson, and Santaquin. Many communities retain substantial amounts of agricultural land embedded in a mosaic of other uses. Near American Fork, wheat fields, orchards, horse pastures, and truck farms are interspersed among residential developments. Many new suburban homes occupy large lots: tractors, vegetable gardens, root cellars, boats, and satellite dishes are commonly seen in backyards. In most areas, old irrigation ditches are still in place and relied on. However, many large agricultural ownerships have been divided into hobby farms and rural residential tracts.

Despite spreading development, Utah County remains a major commercial producer of fruit. Some 81 percent of the state's apples, 60 percent of its sweet and tart cherries, and 51 percent of its peaches come from the benchland orchards east of Utah Lake near Pleasant Grove and Payson. Substantial harvests of pears, apricots, corn, vegetables, hay, winter wheat, barley, oats, chickens, and milk are still recorded.

Utah County Planning Director Jeff Mendenhall ascribes the uneasy coexistence of agriculture and urban growth to the county's long tradition of zoning. Countywide zoning has existed for over 50 years. People seem pretty used to it by now. Agricultural land is not only economically important to county residents, it embodies a lot about who they think they are.

Utah County has several zones which afford at least temporary protection of resources. The A-1 Agricultural Zone does not allow residential use except for dwellings on 40-acre farm units. Most of this land is located outside city boundaries. The Critical Environment Zone (CE-1) restricts houses to one per 50 acres in steep terrain. The CE-2 zone applies to less rugged yet still sensitive lands where one house is allowed on each 20 acres. The Mining and Grazing Zone covers the dry mountain and desert portions of the county where one house is permitted on each 20 acres. Proposals to build houses on land within the Floodplain Overlay Zone trigger a separate set of review criteria related to impacts on water quality, placement of fill, and stream channel alteration.

Mendenhall believes that rural zoning works pretty well as a means of keeping growth within existing cities. The landscape tends to back him up. Ranches up the Spanish Fork at Thistle (site of a massive landslide in 1983) and Birdseye have not yet been split for residential use. Near the LDS

Church's 12,000-acre farm at Elberta and Goshen, the landscape remains agrarian. Genola has been incorporated as a city but remains dominantly an agricultural place. Land speculation schemes away from the main urban corridor have traditionally come and gone, leaving little behind.

The grandest of these development dreams was Mosida-by-the-Lake. In 1908, three promoters purchased 9,500 acres north of Elberta on the west shore of Utah Lake's Goshen Bay. The Mosida Fruit Lands Company split the land into 5-acre parcels, constructed irrigation systems, built an elegant hotel, and marketed the land as peach orchard tracts.

Between 1912 and 1913, some 250 people moved to the remote site after being wined and dined at the resort hotel. Grain fields were planted as an interim crop until the orchards began to bear. However, the project soon unraveled. Only 2,000 of 9,500 acres were sold. Utah Lake rose, lifting water tables into the fields and destroying the irrigation systems which served upland farms. When the alkaline freshwater lake dropped, it left behind a saline crust on croplands. Fields away from the lake proved to be of low fertility. Sieges of grasshoppers devoured grain crops and sent many investors fleeing in frustration.

By 1920, only 67 people remained at Mosida. The community power plant and irrigation pumps were burned out, the fields were full of weeds from sheep overgrazing, and most of the peach trees had been killed by frost. The company went into receivership. The last person hung on in Mosida until 1924. All that remains behind is the foundation of the hotel and ruins of the pump house and power plant. An historical marker stands nearby as a kind of community gravestone. Some of the property has been bought by the Mormon Church and added to its Elberta Farms outfit. The only sound at Mosida today is the clanging of sheep bells on the sage-scented wind.

THE FIGHT FOR UTAH LAKE

Utah Lake is the focus of the county's most intense and far-reaching environmental conflict. Ever since the survey of the "true" lake level by the Federal government in 1885, local landowners have been angry over a perceived taking of their property. They claim ownership of the exposed lake bed and marshlands lying between the survey line and the edge of the water.

The lake's bathymetry makes this matter even more troublesome. The maximum depth of this 100,000-acre water body is 16 feet. When Utah Lake rises or falls a foot, it can mean a difference of thousands of acres to landowners in the flat-lying littoral zone. Some want to use the land for grazing; others envision diking and filling the wetlands and converting them to commercial and industrial sites.

Conservationists have long argued that these habitats should be set aside for wildlife. Provo Bay on the lake's east shore and Goshen Bay at its southern end are fabulously productive waterfowl ecosystems where relatively clean water flows in from the Provo and Spanish Fork rivers. Huge flocks of Canada geese, mallard, pintail, cinnamon teal, northern shoveler, gadwall, double-breasted cormorant, western grebe, green-winged teal, redhead, and other waterfowl depend on Utah Lake as nesting, feeding, and resting grounds. Endangered bald eagles feed on the lake's fish. Birds such as great blue heron, black-crowned night heron, ibis, and snowy egret nest around the lake.

Many of these species are on the National Audubon Society's list of Species of Special Concern. Provo Bay is the only place in Utah where cattle egrets nest. In addition, the marshes support muskrat, mink, kit fox, bobcats, and spotted skunks. Given that the city of Provo is boxed in against the Wasatch and can only expand westward toward Provo Bay, the conflict over Utah Lake has at times been intense.

In 1980, Congressman Gunn McKay introduced the Provo Accretion Lands Bill. Accretion lands are defined as all terrain lying between the 1885 lake level line and the existing edge of the water. The bill, which would have simply given lakeside lands to adjacent property owners, was called "the McKay Steamroller" by conservationists. McKay was clearly supporting development of the wetlands near the Provo Airport. The masquerade that the bill would benefit farmers was unmasked in Washington by the relentless lobbying of the Audubon Society. The bill was defeated in the Senate.

In 1981, the City of Provo introduced a bill in the Utah legislature entitled the Provo Bay Accretion Resolution. The measure was portrayed by Provo lobbyists as a simple, even boring, tidying up of an old land title discrepancy. However, the city's attempts to minimize the importance of the bill backfired. It came to light that Provo's airport was built on accretion land, and the city wanted clear title so expansion could occur into adjacent wetlands. The old city landfill site, also in the accretion zone, was being considered as the site for an industrial park. Lastly, the city wished to connect I-15 with the airport in order to create an industrial district close to air, highway, and rail transportation. Mayor Ferguson of Provo didn't help the cause by admitting that no development would occur in southwest Provo until the wetlands were drained. The bill was defeated by legislators unwilling to give away a resource so prized by the influential duck hunting lobby.

In 1986, Republican Congressman Jim Harrison, a noted fancier of ducks, introduced a bill to create a 30,800-acre national wildlife refuge at Utah Lake. The El Niño floods of 1983 raised the Great Salt Lake to record heights and flooded huge expanses of marshland habitat. The ecological importance of Utah Lake became obvious as massive flocks of displaced waterfowl moved onto its littoral ecosystems. Birders counted over 70,000 ducks on the small water body.

A year earlier, the U.S. Fish and Wildlife Service (USFWS) and the Canadian Wildlife Service set a mutual goal of sustaining 62 million breeding ducks in an average year, some 10 million more than existed at the time. To help meet that target, the USFWS recommended that a 34,000-acre refuge be established at Utah Lake. According to the USFWS, the proposed Utah Lake National Wildlife Refuge would consist of 17,600 acres at Goshen Bay, 12,300 acres at Provo Bay, and 4,100 acres at Benjamin Slough.

The projected cost of buying the contested land was high—$16.5 million. Sobered, the Federal government explored various alternative mixtures of land purchases, conservation easements, management agreements, and leases to protect the resource. However, Utah Senator Jake Garn, still queasy from a NASA junket in outer space, balked at the cost and threatened to use the full force of his office to stop the proposal.

As often happens, a task force was formed to study options for the future use of Utah Lake. Scores of proposals and counterproposals were bounced around. Storage of water from the Central Utah Project in the lake was debated, as were construction of a causeway and the dredging of harbors. Some, like the Utah County Engineers Office, saw the lake simply as a source of water for industry and its waterbirds as nuisance debris clogging the engines of jet aircraft. The causeway concept was eventually dropped by the Bureau of Reclamation, since it would have prevented movement of fish throughout the lake and might have wiped out a population of an endangered fish species known as the June sucker (*Chasmistes liorus*).

Broader concerns about the ecosystem also existed. Agricultural runoff and urban/industrial pollution caused the water of Utah Lake to grow so cloudy, fertile, and warm that it no longer supported trout or mountain whitefish. Fishermen were catching mostly carp, channel catfish, white bass, and walleye.

The entire refuge idea was soon tabled because of a lengthy menu of seemingly intractable issues. For now, all that has been decided is that the State of Utah owns the lake bed. A Provo city planning analyst believes that the city and landowners will eventually win and that in the end, the airport expansion, connecting highway, and industrial park will be built.

Some landowners don't see the issue as that simple. Walking around the muddy marshlands and bottomland forests next to Provo Bay you encounter numerous signs nailed to fenceposts:

Notice to Persons Accessing this Property: Respect Landowners Rights! This land has been private property since U.S. Land Patents were issued on Sept. 30, 1882. Records on file with Utah County Recorders Office.

Near one of these notices, placed amid cottonwoods, ash, and willows, is a far more mysterious sign. It reads: "The Forest of Camelot." Perhaps a landowner was remembering a time before lawyers, politicians, developers, and conservationists took over. Most likely it's the name of a subdivision.

WILDLIFE RESOURCES

The persistence of agricultural land in the Utah Valley protects numerous wildlife species. California quail, ringnecked pheasant, chukar partridge, Hungarian partridge, and band-tailed pigeon are among the upland game-birds which benefit from the existence of open farm and ranch land. Upslope, the foothills and mountainsides of the Wasatch provide critical range for elk, deer, and mountain goats. The Utah State Division of Wildlife Resources (UDWR) has been very active in protecting these ecosystems. Statewide, the UDWR has acquired about 360,000 acres of wildlife habitat, with more than 41,000 of these acres located in Utah County. In Salt Lake County, the agency has run into greater resistance and higher land values—only about 1,900 acres have been bought.

The achievements of UDWR are commendable. In eastern Utah County along Soldier Creek, the Dirty Fork Wildlife Management Area (WMA) embraces 2–1/2 square miles and the Starvation WMA 8 square miles. Near Thistle, the Larson Draw, Jackson, Lake Fork, and Spencer Fork WMAs abut Uinta and Manti–La Sal national forests. UDWR's largest preserve is the 18-square-mile Loafer Mountain WMA east of Payson in Spanish Fork Canyon. This mosaic of Gambel oak/maple brushlands, grasslands, and conifer forests serves not only as critical big game winter range, but as a migration corridor and birthing area for elk and mule deer.

The Utah Division of Wildlife Resources has relied almost exclusively on outright purchases of private land to conserve habitat. In 1983, the University of Utah's Energy Law Center produced a report on conservation easements for the State Department of Natural Resources. This study lauded the potential benefits of incorporating easements into various state agencies' operations.

Ralph Miles was State Lands Director under former Governor Scott Matheson and is an old hand at confronting resource issues in Utah. According to Miles, now the UDWR Land Acquisitions Specialist, conservation easements have not been accepted by state personnel. He argues that since the Division of Wildlife Resources must require public access and removal of livestock grazing, any conservation easement could cost as much as 90 percent of the property's full market value. No wonder the tool isn't being used.

It is unlikely the UDWR will successfully use conservation easements until they are placed in their proper context. To do so, land acquisitions can be divided into three types:

(1) only access needed—best achieved through recreation management agreements or purchase of small streamside access sites,

(2) protection from land subdivision needed—best achieved through a conservation easement, and

(3) full management control needed—best achieved through land donations and direct purchases.

Landowners may be willing to donate or sell conservation easements to the state, but only if they do not *have* to allow open public access and only if they can continue to graze livestock responsibly. The fish and game agencies of other Rocky Mountain states, such as Montana, Wyoming, and Colorado, routinely incorporate easements into their habitat protection programs. Thus far, Utah has not.

SUNDANCE

The Wasatch Mountains in Utah County provide a very different skiing experience from the crowded Salt Lake County portion of the range. Actor Robert Redford's Sundance Ski Resort offers the same superb, dry powder without the attendant crush of condominium development. This small ski area lies next to the Mount Timpanogos Wilderness Area in a setting of supreme beauty. The runs have been cut out of aspen groves and spruce forests in a way that simulates the existing, alternating pattern of closed canopies and open meadows. The only facilities at the base are a small lodge and restaurant (next to "Bob's Pond"), a building where passes are sold, and several meeting rooms.

Redford lives nearby and uses the ski area as a personal retreat and center for his Institute for Resource Management (IRM), Sundance Institute, and film workshops. IRM organizes conferences on environmental issues and acts as a rational intermediary in solving land-use disputes. Redford and the staff of the Institute were instrumental in getting a coal-fired electric generating plant moved off the Kaiparowits Plateau in southern Utah, where the plant would have polluted the air of the Glen Canyon National Recreation Area, the Canyonlands, Arches, and Grand Canyon national parks, and Natural Bridges National Monument. The Intermountain Power Plant, the world's largest coal burning facility, was built instead out in Utah's west desert area near Delta, where its smoke disperses more easily in the winds of the Basin and Range Province.

IRM is involved in environmental troubles across the globe. President Terry Minger reports that the Institute organized a conference in the former Soviet Union on global warming. While thus far focusing his attention on public land issues and worldwide environmental problems, Minger feels that the private market approach to land conservation really matches the IRM philosophy. Redford is himself a calm, clear voice concerning the future of Utah's landscape. He believes that there is still a chance to secure a workable balance of development and conservation, since most of what makes the Wasatch so incredible still exists here.

Redford owns 5,000 acres at Sundance. Yet, only 65 acres have been developed for houses. A maximum of 50 additional acres are set aside for future expansion. While Redford's twenty years in the environmental movement and his formation of the Institute for Resource Management give him solid credentials, in an interview in a 1989 issue of *Outside Magazine*, he spoke about sometimes pulling away from the organization, believing there are times when compromise is unacceptable.

I don't feel I'm tied to the Institute. I can step aside and take a strong position if I feel it, and if it jeopardizes the kind of work we're

Provo: Happy Valley Hangs On

trying to do, then I'm just going to have to deal with that. There are times when I don't feel compromise is the best way to go. At the same time, I believe that by trying to get the opposing sides to work together, ours is a radical approach. (December 1989)

CONCLUSION

On winter days, Provo's Geneva steel mill discharges huge plumes of rust brown smoke into the frigid air. The raunchy clouds scud eastward and sink, enveloping residential neighborhoods. Once, while I was watching this spectacle, a country song came on the radio describing lost farmland and polluted skies caused by baby booms. It was an eerily appropriate soundtrack for the place.

Utah County has long been called Happy Valley due to its smiling, ultra-conservative Mormon population and its extremely high birth rate. The county has nearly doubled in population since 1970 to 255,000. In 1940, there were only 57,000 people there. While thousands of acres have been subdivided and developed, and while the air is often foul, Utah County retains a surprising amount of open space and ecologically important farm and ranch land. The county's zoning regulations are operating as a moderately effective holding action against development in rural areas. Over 60,000 acres of wetlands, 26,500 acres of rangeland, 17,500 acres of cultivated farmland, and 6,000 acres of pasture remain around Utah Lake alone. Unfortunately, Utah County has no land trusts to begin to formalize the protection of these and other crucial lands. For now, zoning and the renowned obedience of the Mormon people will have to suffice.

18

BOUNTIFUL, OGDEN, AND BRIGHAM CITY: SUBURBAN ZION SPREADS NORTH

BOUNTIFUL

When Glacial Lake Bonneville stood at its highest level, the only part of present-day Davis County not to be inundated was the Wasatch Mountains. Today, most of the county still remains covered by the Great Salt Lake. Urbanization of agricultural lands at the foot of the Wasatch spreads from the county line north to Bountiful, Farmington, Layton, and Clearfield. Crude oil is piped in from Wyoming and Uinta Basin fields to Phillips Petroleum and other refineries which soil the sky over North Salt Lake and Bountiful. At night, gas flares burning atop the refineries evoke the fiery wells of war-torn Kuwait. Townhouses, car dealerships, sand and gravel quarries, shopping centers, warehouses, and eight-plex movie theaters parallel nearby I-15.

Bountiful, settled in 1847 by a Mormon named Peregrine Sessions, was once a place where dairy farms, orchards, and fields of sugar beets extended from the foothills to the salt marshes around the lake. Wool and flour mills and a sugar beet processing plant were run by an LDS United Order Cooperative. In 1940, Bountiful had only 3,500 people. Today, it is an urban clot of 45,000. In the film *The Trip to Bountiful*, Geraldine Page portrayed an old woman who wanted to go home to the farming town of her youth. When she finally got there, she found a few weather-beaten buildings and all the people gone. In Bountiful, Utah, the dilemma is more subtle. The town is still there, but it has been transformed into a nondescript segment of the Wasatch Front sprawl.

The foothill zone's deer and elk winter range habitats above Bountiful

∧∧∧∧∧

are being covered by a collection of elite subdivisions. Huge houses back up on Wasatch National Forest lands. The high "Bonneville"-level glacial lake shoreline marks the most expensive neighborhoods. Local kids go to the Oak Hills School and an elegant new LDS ward house. The "Provo" shoreline terrain downslope is more middle class. The flats along the interstate contain a clutter of trailer parks, storage lockers, processing plants, and burger stands. Only the giant Pillsbury grain elevator complex at Kaysville, which in the distance looks like an LDS temple, indicates this was once farm country. Today, most of the grain processed at Pillsbury comes from the north in Box Elder County and the Snake River Plain of Idaho.

There are no formal open space conservation programs in the county seat of Farmington, in Bountiful, or in any portion of Davis County. Some orchards are still found around Kaysville, but the agrarian pattern has largely disappeared. The county's most popular recreation attraction is the Lagoon Amusement Park and Pioneer Village next to I-15 near Farmington. Lagoon is noted for its miniature golf course, roller coaster, swimming pool, zoo, rodeos, demolition derbies, and re-created pioneer farming town.

The most important ecological and open space resources in Davis County are associated with the Great Salt Lake. Although Kit Carson once explored the area, it was Howard Stansbury of the U.S. Army Corps of Topographical Engineers who first studied the lake systematically. In 1849 and 1850, Stansbury surveyed its shoreline and investigated its islands. He reported seeing massive flocks of great blue herons, white pelicans, cormorants, and gulls and marveled at the sun-blotting clouds of other waterfowl and shorebirds. Stansbury was awed by the strange elegance of the saline wilderness he moved through. He wrote, "All is stillness and solitude profound."

Stansbury named the lake's largest island after the herds of pronghorn antelope which inhabited its grassy slopes. The Mormon leadership quickly saw the merits of Antelope Island as livestock pasturage isolated from wolf and coyote predation. Kaysville residents soon stocked the island's fragile prairies with over 10,000 sheep. The range became devastated.

In 1877, Adam Peterson bought 10,000 acres on the island, moved the sheep off, and shot the wild horses which roamed the rocky terrain. In 1892, William Glassman brought 12 buffalo to the island as an experiment and tourist attraction for a beachside resort. This herd eventually became naturalized and has since expanded to 400 head. The buffalo were privately managed until 1981, when the State of Utah acquired the entire 27,000-acre island and created Great Salt Lake State Park.

Utah: An Elusive Zion

Antelope Island had been held by the Basque-owned Anschutz Corporation, which now controls the Denver and Rio Grande Railroad. So coveted were the island's ecological, open space, and recreation resources that the state resorted to legal condemnation proceedings in order to buy the land. The causeway road linking the island with the mainland was flooded by the rising lake in 1983 and has remained underwater ever since. This is nothing new. Over the years, as the lake moved up and down, this alternately allowed and blocked the movement of game on and off Antelope Island. During the droughts of the 1930s and early 1960s, it was possible to walk to the "island" from Bountiful. Today, although the lake averages 14 feet deep, the only access is by boat or helicopter. For now, the buffalo and other wildlife live largely undisturbed.

The fertile marshes around the Great Salt Lake in Davis County are part of one of America's most productive avian ecosystems. Approximately 30 percent of all ducks and geese in the Pacific and Central flyways converge on these wetlands during fall and spring migrations. Some 3 million ducks, 1 million phalaropes, and 70,000 California gulls depend on these complex littoral environments where rivers and groundwater flows enter a saline, aquatic desert. Some 80 percent of the world's white-faced ibis nest in the Layton–Farmington Bay region. Three-quarters of the Pacific Flyway's annual production of whistling swans are also dependent on these environments. In addition, snowy plovers, egrets, white pelicans, sandhill cranes, avocets, black-necked stilts, long-billed curlews, and wintering bald eagles rely on what many consider salty wastelands for their survival. In Davis County, the main protection of these critical ecosystems has thus far been within the state's Farmington Bay Waterfowl Management Area.

In 1982, The Nature Conservancy (TNC) became involved in the effort to protect the Davis County marshlands. TNC acquired an option to buy 1,192 acres from the aerospace firm Morton-Thiokol in an area known as Layton Marsh. A year later, the Great Salt Lake stood at record heights, and feeding and nesting areas within state and Federal refuges were flooded. Slightly inland, Layton Marsh served as emergency habitat for tens of thousands of birds. The Utah Division of Wildlife Resources announced, "With State refuges like Ogden Bay and Farmington Bay sustaining considerable disruption, it is obligatory that buffer areas like Layton Marsh be secured" (The Nature Conservancy, press release, December 1, 1983).

In June 1983, Secretary of Interior James Watt wrote Utah Senator Jake Garn indicating that "permanent protection of this property [Layton Marsh] by The Nature Conservancy through a gift or bargain sale is consistent with the private sector emphasis of my new wetlands task force" (June 8, 1983, in files of The Nature Conservancy). TNC and the Utah Wildlife Habitat Development Foundation began a joint fundraising effort around the state. Later that same year, a deal was struck with Morton-Thiokol. The corporation agreed to sell 1,192 acres to TNC for a bargain sale price of $325,000, far below the fair market value. The corporation could then write off the difference as a tax deductible charitable gift. When administrative overhead, legal fees, and fencing costs were added, Layton Marsh was a $500,000 project.

Unfortunately, Layton Marsh is an isolated case in Davis County. As Governor Bangerter's pumping facility at Lakeside continues to lower the Great Salt Lake, thousands of acres of marsh habitat are drying out and attracting the attention of developers in search of industrial and commercial land to fill and sell. For the present, though, private duck hunting clubs own and manage the majority of wetland sites for wildlife.

While this manner of protection is welcomed, it should not be relied on in the long term. If the number of ducks and geese declines as a result of the destruction of other habitats, the hunting clubs may sell to developers. The formation of a Great Salt Lake land trust seems the most viable way to conserve these sensitive ecosystems. Such a trust could draw together the duck clubs, independent duck hunters, farmers, industrial landowners, and conservationists to begin to save the extensive wetlands which remain.

OGDEN

Utah's dependency on defense contracts makes Utah's state economist, Randy Rogers, nervous. He believes that the current cuts in defense spending will badly hurt the state's economy. Ogden was once a major railroad center and junction point on the transcontinental railroad. Named for Hudson's Bay Company fur trader Peter Skene Ogden, the city is now a manufacturing center for the defense and space industries. Hill Air Force Base alone employs 22,000 civilians. Aerospace and weapons giant Morton-Thiokol is the largest private employer in Utah with 8,000 people on its payroll. Its steel-and-glass headquarters building is adjacent to the LDS temple in downtown Ogden. TRW, Browning, and other defense contractors also operate plants in this vest pocket county of 165,000 people. Wallace Stegner wrote of the original gathering of Zion as an agrarian, utopian movement. University of Utah historian James Clayton has called the state's dependency on armaments manufacture as "an unhallowed gathering." A Weber County official puts it more plainly. "If Hill Air Force Base closes and the defense industries leave, Ogden and Weber County would cease to exist."

Defense spending constantly injects huge sums into the economy of the Ogden area. The city is undertaking a downtown revitalization project, a housing rehabilitation program is in place, and a Hilton Hotel and shopping mall have been built. Jobs at the IRS Western Internal Revenue Center add to the region's massive Federal economy. Other employers include Utah Power and Light, the Kier Corporation (a powerful housing contractor), the Amalgamated Sugar Company (which processes sugar beets), Pillsbury, banks, and retail businesses. Perhaps the most unusual company in Weber County is Living Scriptures. This corporation creates a series of cassette tapes such as "Dramatized LDS Church History" and "The Dramatized Bible" in its Ogden production studios. As a result of this base of jobs, city planners perceive the area as "pretty much built-out."

Population figures support this claim. In 1960, the city of Ogden had 71,000 people. Today, as the surrounding agricultural land has become sub-urbanized, it has only 64,000. Developments such as Forest Green Estates, Shadow Valley, and Eyrie Meadows have consumed orchards and pastures. Much of the benchland at the foot of the Wasatch has been carved into residential neighborhoods such as College Heights near Weber State College. However, enough dairy farmers remain in Weber and adjacent counties to support the large Cream O'Weber dairy.

Like most Wasatch Front cities, Ogden is backed up against the mountains. The Sensitive Area (SA) Zone was created along the terrace of the Wasatch fault to coincide with the pattern of open space and wildlife habitats. The regulations for the SA zone require rigorous analysis of proposed developments both along the steeply sloping toe of the mountains and within riparian corridors. In places, it seems to be working.

The view from "Sensitive Area"–classified foothill lands above Ogden sweeps across most of Weber County. To the north, houses are spreading up the Lake Bonneville benches into irrigated orchards below the Ogden-Brigham Canal. South, toward Weber State College, developments will soon merge to form a full grid. Farther south, the sky is crowded with fighter jets above Hill Air Force Base. Westward, the flat pastures and marshlands have been extensively broken into 5- to 40-acre tracts. Horses, cattle, and sheep graze next to trailers, homemade houses, and broods of children. Beyond lies Fremont Island, partially obscured by reflected sunlight and salt haze.

For now, the privately owned benches above Ogden are serving as vernacular parks next to the grasslands and oak thickets of the Wasatch National Forest. Children ride bouncing mountain bikes over the eroding deltaic sediments, and adults walk their dogs among the oaks and sagebrush. At sunset, the sky and distant lake surface light up to form a continuous arc of shifting colors.

PLAIN CITY

Mormon people envision even the smallest settlement as a potential metropolis. All over Utah, tiny hamlets are called cities. Places such as Bear River City, Garden City, Heber City, Oak City, and Spring City are actually pleasant, insular small towns. Brigham Young's colonization drive was full of expansive optimism. In the fervor of the era, people believed that with the help of God, every colony would be built into a grand urban center. Therefore, it would be a failure of faith to call even a nascent farming settlement a town. After all, St. Augustine spoke of the *City* of God, and Joseph Smith planned a righteous *City* of Zion. Towns are for nonbelievers.

Utah law acknowledges the Mormon's faith in growth. Section 10-2-301 of the Utah Code Annotated defines first class cities as 100,000 people and above, second class cities as 60,000 or more, and third class cities as those with at least 800 residents. However, the law also states, "This section shall not lower the class of a municipality which now exists." As a result, numerous one-store, one-LDS ward towns settled in the nineteenth and early twentieth centuries are grandfathered in as true cities.

Plain City is a classic representative of this social and economic idealism. By 1859, Mormon settlers had begun farming the flat-lying plains at the edge of the Great Salt Lake northwest of Ogden. Walking around the community, you sense that this is still a place people count on. Enough farms remain to fill the November air with the sweetness of corn silage and cow manure. The smell of pastures being burned near the lake carries a strange saltiness. The sound of bawling holsteins seems to be answered by the honking of Canada geese flying in "V" formations overhead. Sonic booms from Air Force jets occasionally rattle the serenity.

Suburban homes are being built around Plain City as farmers sell road frontage in an effort to keep their operations going. The town is slowly being converted to a rural bedroom community. Yet, the infill of newer homes is patchy. Holland hay rakes are parked next to pastel blue and green tract houses. Traditionally, Mormons have kept their farm implements in town, but this is an intriguing variation. Blond children dressed in the latest lavender clothing and Air Jordan basketball shoes use the rakes like jungle gyms.

Plain City is a glorious mixture of material culture. A horse or pony grazes in back of a typical house next to a livestock trailer, pickup camper, tractor, and satellite dish. Freshly planted lombardy poplars grow in rows along property lines. At Halloween, hay bales, corn stalks, and pumpkins are arranged on the doorsteps of tract homes which have been built next to working farms where these products are anything but ornamental.

Down one quiet street, a row of newer homes stands on a low terrace above the saltmarsh. A trio of Big Wheel tricycles sit to one side of a small patch of alfalfa. Down below is the sound of children playing. A well-worn path leads away into the marshlands, pastures, and saline mudflats. Nearer the lake, the depth to brackish and fresh groundwater varies in an indecipherable way. Inland saltgrass, sheep fescue, shadscale, spiny hopsage, greasewood, galleta, Indian ricegrass, bottlebrush squirreltail, rabbitbrush, broom snakeweed, sego lily, prickly lettuce, Russian thistle, pickleweed, pepperweed, fleabane, and tamarisk grow in patches and swirling patterns. Brazenly polygamous male red-winged blackbirds sit atop cattails flashing epaulets and defending their turf with loud "checks" and calls of "honkareee."

Above the terrace scarp, an old man putts on a homemade green in his backyard. He offers, "I've been here most of my life, and the change isn't so bad." Pausing to drop a well-worn Titleist into the front pocket of his brown jacket, he concludes, "They're good local people, and we welcome them to add to our community and build it up a bit."

Plain City is reminiscent of a small New England town. At its center, there is a park surrounded by an LDS ward, stores, a school, and older white homes. While the AG Food Store has gone out of business, Plain City Drug, Plain City Confectionery, Richard's Service Station, and Country Cupboard Cafe seem to be doing well. There is even a new credit union. At Jack's Garage, a melange of horse trailers, lawn mowers, tillers, tractors, and RCA televisions are sold.

Basketball hoops seem to be everywhere. Plain City could also pass for an Indiana farmtown where boys dream of winning the state championship and playing for Bobby Knight. Here, though, the dream is to represent not only your state but your faith to the rest of the world. The boys shooting hoops beneath the cottonwoods will no doubt go on two-year Mormon missions. But the fire in their eyes is to play for Brigham Young University. They sink jumpshot after jumpshot. Perhaps one of them will be the next

Danny Ainge or Michael Smith and wear the kelly green of the Boston Celtics. It is a community long on fundamentals with a calm and justifiable pride. While there is no formal land protection program there, the cohesiveness of the people and their attention to the basics is conserving much of what is important.

Weber County planners believe that despite so much development, there is no "open space problem." Given that much of the area has been urbanized and there is seemingly little concern for conservation of what is left, it is unlikely that very much of the remaining orchards, pastures, marshes, river corridors, and big game habitats of Weber County will survive. The area's reliance on the defense and space industries for employment has resulted in a substantial reduction in agriculture and has detached many residents from daily connection to the natural environment. While pockets of sustainably used land remain, such as Plain City, the county careens along undeterred by the losses.

BRIGHAM CITY

Brigham City is as well-named as Coal City, Illinois, Lumber City, Georgia, and Highway City, California. While originally called Box Elder, this hard-working, middle-class community was renamed in 1856 to honor the Mormon economic patriarch Brigham Young. It was the center of the Mormons' most successful United Order Chapter, known as the Brigham City Mercantile and Manufacturing Association. There was a cooperative store, tannery, wool factory, livestock yard, sawmill, public works department, flax farm, and foundry—in all some forty divisions. Local people held stock in the association and were paid dividends when times were good. However, following a fire at the wool factory, loss of crops to drought and grasshoppers, and the growth of competing businesses, the divisions were sold off one by one to private interests. By 1895, the cooperative store was bankrupt.

Brigham City is now a comfortable county seat of about 18,000 residents spread out at the base of the Wasatch. It is still known as the Peach City, since substantial acreage of orchards exists south of town on benchlands above the lake. U.S. Route 89, which traverses the area, is called the Fruitway. Truck farms, livestock operations, and grain fields cover much of the county landscape. Many of these properties have been in continuous use since the early days of the Desert Land Act and Enlarged Homestead Act.

Land development pressures in Box Elder County are not nearly as intense as those in the rest of the Wasatch Front. The county planning program concerns itself mostly with routine matters of zoning and subdivision review. The Planners and Engineers offices have been combined—a classic nuts and bolts arrangement. The only conservation projects near the city are the state's Brigham Face Big Game Wildlife Management Area on the slopes of the Wasatch. However, in downtown Brigham City, twin iron girders support a large neon sign which spans Main Street. This landmark boldly crows, "Welcome To BRIGHAM, Gateway—World's Greatest Game Bird Refuge."

Outside the city, along the east and north shorelines of the Great Salt Lake, are some of the most important waterfowl and shorebird habitats in the United States. The 65,000-acre Bear River Migratory Bird Refuge occupies the Bear River delta and adjacent bays. This vital Federal refuge is used by ducks, geese, herons, cranes, and many other birds from both the Pacific and Central flyways. The U.S. Fish and Wildlife Service purchased the land in 1928 to halt species decimation from overhunting and the draining of wetlands. During the 1890s, commercial hunters were selling mallards from the Great Salt Lake marshlands for $1.50 a dozen. One report described how a single hunter killed 325 ducks in one afternoon at the lake. With the establishment of the refuge, some sixty species now nest in peace.

Although hunting continues in the ecosystem, rational bag limits and seasons have been set. Mounting the 100-foot-tall observation tower at the refuge during spring and fall migrations, it is possible to observe a startling diversity and quantity of birds. Documented sightings of pink flamingos have even occurred. This habitat is on a par with the Stillwater and Ruby complexes in Nevada, the Platte River in Nebraska, and the Bosque del Apache refuge on the Rio Grande in New Mexico.

Upstream on the Bear River toward Tremonton is the state's Salt Creek Waterfowl Management Area. The Willard Bay Reservoir along the lake's east shore is a freshwater impoundment built in 1964 by the Bureau of Reclamation. Located next to I-15, this 15-mile-long diked pond stores water from the Weber and Ogden rivers during the winter months for summer cropland irrigation. Its construction destroyed excellent natural waterfowl habitat, but wildlife still make use of the reservoir itself. However, the water body is also extremely popular with motorboat jockeys.

West of Brigham City is the now-quiet farming town of Corinne. During the construction of the transcontinental railroad in the late 1860s, Corinne was the most Gentile community in Utah. Chinese, Polish, Czech, and other peoples from all over the world camped there when it was a wide open frontier boom town. The so-called "Burg on the Bear" once had a population of 2,000 and supported a score of bars and bawdy houses. In 1870, the town also had Utah's only Presbyterian and Methodist churches. An historical marker in the city park reads, "the educational and religious impact cannot be measured," leaving the reader unsure of whether this means it was large or inconsequential in dominantly Mormon Utah.

After completion of the transcontinental railroad in 1869, most of Corinne's residents dispersed, leaving the area open for reoccupancy. With construction of the Bear River Canal, the sagebrush grasslands began to be farmed by Mormons. For a while, a steamboat named the *City of Corinne* cruised the lake from a beach resort at the mouth of the Bear River.

Corinne is now home to about 500 Mormons who support two LDS wards. The street signs reveal the city's mixed history. Each bears a name, such as Colorado, Arizona, or Montana street, with 2300 North or 1300 East shown in parentheses beneath. The Mormon numerical grid has been superimposed on the Gentile naming scheme. The downtown contains the Bear River Valley Cenex Co-op grain elevators, the Sunshine Elevator Company (which advertises "Custom Rolling"), feed and grain shops, and Mim's Cafe.

Just west of Corinne on Route 83 stands a curious road sign. There, out in the quiet of grainfields and sagebrush, the sign warns, "Morton-Thiokol Shift Change 3:30–4:45, Please Use Headlights!" If you are heading west during that part of the day, you are met by a steady stream of cars and pickups—about thirty per minute. For behind the croplands, silos, and sheep, Box Elder County has an unexpected alter-ego.

Twenty-five miles west of Brigham City, in a low spread of brown grasslands called the Blue Hills, lies Morton-Thiokol's sprawling test areas and production plants. A sign at the guarded entrance calls it "The World's Largest Facility for Solid Rocket Motor Research and Production." Formed in 1982 by the merger of the Thiokol Chemical Company and Morton Salt, Morton-Thiokol has become a behemoth space and defense contractor. Its Space Division alone covers 20,000 acres. Solid fuel rocket motors for the NASA Space Shuttle are built and tested there in the vast solitude of open space. The room is needed. Each motor contains 1.1 million pounds of propellant.

Since the disastrous *Challenger* explosion, Morton-Thiokol has staged a public relations rebound. The company's Strategic Division makes solid propellant motors for nuclear ballistic missiles such as the "Peacekeeper" and various weapons carried by the Navy's Trident II submarines. The division is also working on the Midgetman ICBM and the B-1B bomber. The Tactical Division produces single and multiple burn rocket motors for air- and surface-launched missiles such as the Navy's MK-104 and the HARM antiradar missiles.

Morton-Thiokol land holdings are scattered widely in the Great Salt Lake region. They include the Morton Salt evaporation ponds near Saltair and the Magna stack, numerous lakeside properties, and the huge rocket motor test complex. Thus far, the company's single conservation effort has been the Layton Marsh project. However, the majority of its land already serves as *de facto* preserves. Despite occasional rocket blasts, the corporation's Blue Spring Hills property is excellent habitat for upland game birds, raptors, and big game. As in the Bacchus Missile Works near Magna, the required safety buffers provide significant ancillary support for wildlife. Morton-Thiokol's north shore marshlands supplement the State Division of Wildlife Resources' Public Shooting Grounds and the Locomotive Springs Waterfowl Management Area, the U.S. Fish and Wildlife Service's Locomotive Springs Refuge, and the many pieces of habitat owned by private bird hunting clubs.

PROMONTORY MOUNTAINS

The nearby Promontory Mountains jut into the Great Salt Lake like Italy into the Mediterranean. At the southern tip of the range, the Southern Pacific Railroad's causeway, known as Lucin Cutoff, touches land for a short stretch before heading west across the water. Watching a freight train seem to float across the water is one of the West's truly bizarre spectacles.

North of the peninsula is the most famous site in American railroading. On May 10, 1869, the Union Pacific and Central Pacific met to complete the country's first transcontinental railroad. On that sunny day out in the sagebrush, two golden spikes were driven to complete the link-up. For good measure, a silver spike from Nevada, an iron, silver, and gold spike from Arizona, a silver spike from Idaho, and a gold spike from Montana were also pounded ceremoniously into a special tie of polished California laurel. A 2,200-acre area is now set aside as the Golden Spike National Historic Site.

Historian Leonard Arrington called completion of the cross-country railway "a catalyst, an exciter, a pump primer which speeded up the process of settlement and escalated the West's income" (Arrington, "The Transcontinental Railroad," 15). The space shuttle solid fuel motors built a few miles from the scene of the golden spike have served much the same function. It is difficult to imagine that this isolated place has played a central role in the geography of transportation.

Early winter is the best time to visit Promontory. The 1870s vintage tourist trains have stopped running and an appropriate quiet descends. At dusk, the view east sweeps across grainfields, grasslands, marshes, mudflats, pools, and the lake, all alive with subtle lengthening light. Fremont and Antelope islands appear golden and mythical in the distance. In the growing darkness, the narrow string of settlements along the Wasatch Front begins to glow as lights are switched on while dinners get underway. Above the cities, the white summits of the Wasatch are filling with snow. Offshore is the chokingly saline water of the Great Salt Lake. In between, on the terraces, benches, and deltas, is the hospitable terrain where settlement succeeded. The band of illumination from tens of thousands of houses illustrates more clearly than any map that the Mormons were and continue to be an oasis culture.

The hold of Mormon pioneers was tenuous. Now as agriculture passes away and an urban-industrial way of life takes its place, the Mormons' grip on the Wasatch Front remains as uncertain as ever. Severe problems exist with water quality and supply, air pollution, and habitat loss. The rate of population growth in areas such as Provo exceeds Rwanda. Yet precious few steps have been taken to secure the future of the Front. In the clean air and wild, open stillness of the Promontory Mountains, it is tempting to dismiss the urban corridor as a lost cause. After all, with the Wasatch Mountains, the marshlands, and the West Desert to escape to, why worry about the Front?

Tremendous open space, ecological, recreational, and historic resources remain within the urban conglomeration. It is not too late to protect or restore natural remnants of Zion. The residents of the Wasatch Front, not The Nature Conservancy or the Trust for Public Land, bear the ultimate responsibility to act. Through the formation of local land trusts, involvement in city and county planning programs, and support of state land acquisition programs, much could still be done. Will the legendary work ethic and energy of the Mormon people be turned toward these tasks? Only they can answer.

19

THE WASATCH BACK-VALLEYS: MORMON HEARTLAND

The striking north-south structural trend of the Wasatch Range broadly controlled the pattern of Mormon pioneer settlement. Arable and developable lands were initially claimed in the wide gently sloping Salt Lake and Utah valleys lying west of the glaciated Rocky Mountain ridgeline. From 1847 to 1856, Mormons built an elongate belt of "fall-line" cities where westward flowing rivers and creeks emerged from narrow Wasatch canyons. By the early 1860s, the primary phase of development crested, and a wave of settlement broke backward into a parallel echelon of valleys lying east of the main range. Geographer Donald Meinig has called these small, high basins the "Second Tier" of Mormon colonization. Fed by water from the surrounding mountains and high plateaus, these narrow landscapes were transformed by resolute Mormon labor into some of the West's most productive farms and ranches.

In recent decades, as the Wasatch Front became urbanized, land development mirrored this same pattern and began to flood eastward into the Wasatch back-valleys. Growing numbers of people are now finding respite from the smog and urban congestion of Salt Lake and Provo by recreating in or moving to Wasatch, Summit, and Sanpete counties. Between 1970 and 1990, Sanpete County grew by 51 percent to 16,259 residents, Wasatch County by 74 percent to 10,089, and Summit County by 168 percent to 15,518. This sudden burst of population has begun to dissolve the long-standing ecological and cultural isolation of the back-valley country.

∧∧∧∧∧

WASATCH COUNTY

Wasatch County, like Ouray, Colorado, calls itself the "Switzerland of America." Mount Timpanogos and other spectacular glaciated peaks rise west of Heber City. Named for Mormon pioneer Heber Kimball, the county seat is home to some 4,500 conservative farmers, ranchers, and retailers. The community is disarmingly friendly. Doves Happy Food and Drug even has a sign out front saying, "Lower Prices: Because We Like You!" Many locals are descendants of Swiss Mormon settlers who raise high quality livestock in pastures cut from oak brushlands and precisely irrigated by networks of canals and ditches.

In the 1930s, Wasatch County was a major rail shipment point for sheep and other agricultural products. By the early 1970s, trucking had captured the valley's freight business and an 18-mile-long portion of the rail line from Heber City to Provo Canyon was converted to a tourist train known as the Heber Creeper. A tidy re-creation of a pioneer village serves as the depot. It's Heber's biggest draw.

West of Heber City is the 22,000-acre Wasatch Mountain State Park. This, the largest of Utah's state parks, contains a wealth of forest and brushland ecosystems within its benches and mountain slopes. A twenty-seven-hole golf course and large campground have been built at the main entrance. Land subdivision pressures are growing near the park. The 868-unit Mayflower Resort Community is under construction 8 miles north of Heber. In recent years, Midway has changed from a farming hamlet to a rural suburb of Provo. Oak Haven subdivision, Midway Farms, and Swiss Mountain Estates sprawl over an old geyser basin.

The tentacles of the Central Utah Project (CUP) extend throughout Wasatch County. Just north of Heber City in the Provo River narrows, the Bureau of Reclamation is building Jordanelle Dam. According to Bureau literature:

> *The Jordanelle Dam and Reservoir will store excess flows of the Provo River and other flows normally stored in Utah Lake. Utah Lake storage water will be replaced with water diverted from Strawberry Reservoir and from project return flows. . . . In addition . . . 10,000 acre feet of storage will be allocated to Provo City. Municipal and Industrial water stored in Jordanelle Reservoir will also be delivered to Salt Lake County. (Bureau of Reclamation, 2)*

Utah: An Elusive Zion

The 300-foot-high dam will back a 320,000-acre-foot reservoir up onto some 3,100 acres of riparian bottomland and adjacent grasslands. A 1,450-acre subdivision has been platted next to the site in anticipation of the reservoir's recreational appeal.

During the late 1980s, when the CUP was slated to be reauthorized by Congress, political infighting broke out within Utah's congressional delegation over the cost and impacts of the system. The reauthorization process was used as an opportunity to recommend changes in the shape and price of the undertaking. Democratic Representative Wayne Owens proposed that the Bureau of Reclamation establish a wildlife protection trust fund of up to $75 million. Owens called the CUP "a massive manipulation of Nature" and added, "We are foolish if we believe we can project in advance its total environmental consequence" (*Salt Lake Tribune*, June 10, 1988).

Various proposals were debated, including the restoration of wetlands around the Great Salt Lake and the purchase of vast acreage around Utah Lake. Off-site mitigation funds had already been spent to "replace" habitat lost to reservoirs such as Jordanelle. The Utah delegation agreed that some continued funding would be attached to the reauthorization request. In truth, under Section 8 of the Colorado River Storage Act, the Bureau of Reclamation is required to mitigate habitat losses occurring as a result of dams it builds. In 1986, $4.6 million was budgeted, but only $1.4 million was spent on habitat protection. The rest was redirected into dam construction. Utah's environmentalists howled in protest. The Nature Conservancy held its breath as the debate prattled on. One of its most important projects in the state depended on the availability of increased off-site mitigation funds.

STRAWBERRY RIVER

The waters of the Strawberry River naturally flow eastward into the Uinta Basin. There they join the Duchesne River, the Green River, and eventually the Colorado in Canyonlands National Park. However, a case of cultural stream capture engineered by the Bureau of Reclamation causes a great deal of the Strawberry River to be stored in a reservoir for piping to the Wasatch Front. By the 1980s, despite years of abuse, the Strawberry River remained one of Utah's best brown and cutthroat trout streams and riparian corridors. In 1988, Bill Geer, Director of the Utah Division of Wildlife Resources (UDWR), said in a press release:

The Wasatch Back-Valleys

*We've been interested in protecting the Strawberry River downstream
of the reservoir for twenty years. It's an extremely important piece of
river habitat within a short drive of the Wasatch Front. Statewide, only
about 15 percent of the streams are closed to public access. But near the
Front, the public has lost resources and access on 40 percent of the stream
reaches and this is increasing rapidly. (Utah Division of Wildlife Resources, press release, October 14, 1988)*

Back in 1986, The Nature Conservancy (TNC), the UDWR, and the Bureau
of Reclamation had begun a creative, cooperative effort aimed at protecting
a key 18-mile stretch of the Strawberry from subdivision and dewatering.

In 1987, TNC purchased 700 acres along the river for $1.1 million. The
land was owned by the Dynamic American Corporation's Camelot Resort
Company, which had proposed a one-hundred-lot subdivision on the parcel.
TNC also secured an option to buy an additional 1,730 acres and began negotiations for another 640 acres. The Conservancy anticipated that the Bureau of Reclamation would secure the off-site mitigation money needed to
repay the outlay for the 700-acre piece and step in to buy the other land
before the option expired. True to its word, the Bureau did exactly that.

On October 14, 1988, a reception was held at the Strawberry Pinnacles
beside the river. At this dedication ceremony, it was announced that TNC
would temporarily hold the 3,070 acres until UDWR completed its management plan. The land would then be transferred to the state as a wildlife
management area.

However, the Strawberry River Project is only half done. It was conceived as an 18-mile corridor. Ten miles remain unprotected. Many believe
the Bureau of Reclamation should purchase these lands to fulfill the agency's
mandate to mitigate the impacts of its actions. The Bureau disagrees and has
proposed instead to buy public access easements along that reach. Given the
scope of the environmental destruction directly caused by the CUP and the
urban and rural growth it waters, a request to purchase the rest of the Strawberry seems reasonable. Utah has few good trout streams left.

Full protection of the Strawberry River would establish a unique conservation area. Some 42,000 acres of big game and fisheries habitat near the
river have already been purchased by the UDWR. The nearby High Uintas
Wilderness Area caps the massive east-west range along the Utah/Wyoming
border. The Uinta Basin, in adjacent Duchesne and Uintah counties, is remote sagebrush country with a history of white treachery against Indians,

failed land speculation schemes, and battles over oil and gas development. To its credit, the UDWR has purchased some 97,000 acres of big game, upland game birds, fisheries, and waterfowl habitat in these two counties. Yet, despite all the achievements, the Strawberry—the life's blood of the region—remains unsecured.

SUMMIT COUNTY

Echo Canyon is an eerie, narrow stream-cut slot in eastern Summit County traversed by I-80, the Union Pacific Railroad, and several utility lines. The high sandstone cliffs which rise abruptly on both sides were once used by Mormon pioneers as defensive fortifications against what they saw as an "invading" U.S. Army. In 1857, the Federal government dispatched troops to Utah to quash Brigham Young's attempt at forming the separate, autonomous nation of Deseret. In anticipation of sniping at the unsuspecting cavalry, battlements were built in the sandstone escarpments overlooking the canyon. Fortunately, negotiations prevented bloodshed. Today, the Utah Highway Department's sign marking the site of these fortifications still conveys lingering acrimony.

> *In 1857 due to false official reports and other misrepresentations, troops under General Albert Johnston were sent to suppress a mythical rebellion among the Mormons. Brigham Young, the Governor, exercising Constitutional rights, forbade the army to enter Utah on the grounds that there was no rebellion and that he had not been officially informed of the government's action.*

Perhaps one of the reasons for the State's choice of wording is that Echo Canyon was also the route of the Mormon Trail into Utah. It is hallowed ground. In one stretch near Henefer, a monument containing concrete models of the Book of Mormon, Doctrine and Covenants, the Pearl of Great Price, and the Bible is topped with a large sculptured beehive.

Although recognized as a National Historic Trail, there has been little action to protect the Mormon Trail corridor from development. Plaques alone do not conserve history; it requires land. The LDS Church would be the logical body to establish a Mormon Trail Land Trust which could secure conservation easements from landowners in order to maintain significant portions of the route. Farmland, scenic open space, and wildlife habitat could

The Wasatch Back-Valleys

also be protected along with the historic trail. At present, not even the sign-posts along the trail are being maintained.

Nearby Henefer, Utah, is a town so quiet you could hear a soul being saved. North of town, where the Weber River flows past Devil's Slide, is a winter concentration area where bald eagles peacefully feed on trout. Golden eagles, ferruginous hawks, and red-tails nest in nearby forests. Big game reach record size in the area's fertile limestone habitats. In the 1970s, this wildlife diversity and scenic solitude attracted outsiders with thick wallets. A Taiwanese man bought a ranch and charged up to $3,000 to hunt for trophy bull elk. In 1978, a company called Great American Realtors began advertising the Eagle Ranch Preserve. Its marketing pamphlet explained:

> *You purchase a deeded interest in the Preserve. This deed entitles you*
> *to recreational use of the entire 8,300 acre property. At the same time,*
> *the land is protected from subdivisions as well as State Park despoliation.*
> *You'll share this rugged wilderness with the living symbol of our nation —*
> *the bald eagle.*

Developers promised swimming pools, tennis courts, an indoor/outdoor equestrian area, hiking and riding trails, and an RV campground. The Eagle Ranch Preserve was essentially conceived as a cross between an outdoor health club and a time-share ecosystem.

Problems arose. People were unwilling to pay $8,000 to become members of the Eagle Ranch Preserve Owners Association when most of the same recreational experiences could be had free on public lands or city parks. Many didn't like the confusing joint ownership aspect of the enterprise. Prospective buyers would often ask, "But where's *my* piece of the ranch going to be?" Trepidation increased when the development of facilities was repeatedly deferred to a mythical "Phase Two." The company offered no dependable figures on how many memberships would be sold or how much future maintenance fees might be. There also never seemed to be a clear explanation offered of how the undeveloped land would be protected from subdivision. Conservation easements, deed restrictions, or other tangible legal measures were never mentioned. Suspicion and disenchantment haunted the undertaking from its inception.

In 1982, the Eagle Ranch Preserve went broke and was placed in receivership. The UDWR immediately expressed interest in acquiring the property, since it already owned a 4,000-acre wildlife management area north of the ranch which served as critical winter range for the large Lost Creek mule deer herd. Over 1,000 deer were known to winter on the Eagle Ranch Preserve, along with 60 bald eagles which wintered in the ranch's cottonwood bottomlands along the Weber River. Sage grouse leks dotted the uplands. The UDWR saw the ranch as a high priority link between its 4,000-acre site and the Echo Canyon raptor preserve. A few years later, with technical assistance from TNC, the state bought 7,000 acres of the 8,300-acre property and declared it the Eagle Preserve Wildlife Management Area. At last the name fit.

Land and beliefs are tightly held in the area south of Echo Canyon known as Kamas Valley. In the town of Marion is Utah's most famous LDS ward building. "Keep Out!" signs still hang from chains strung across entrances to the ward's lot. In 1987, members of the polygamous Singer-Swapp clan firebombed their own chapel. They believed that such an act would resurrect their dead patriarch who would return Utah to its agrarian and fundamentalist roots. In their own way, they were rebelling against the state's unrelenting urbanization.

After the bombing, an armed siege broke out at the clan's ranch as law enforcement personnel descended on Marion. During the firefight, a deputy was shot to death. After a five-day standoff, the clan was finally arrested. Some estimate that between 10,000 and 50,000 people in Utah live in polygamous households. While illegal, the practice is usually ignored. The Singer-Swapp episode is a bizarre anomaly. The overwhelming majority of Utah's "polys" are quiet farming families steadfastly hanging on to old ways, and a central part of those odd and alien beliefs is to live close to the land and not give it up.

Jerry Smith, the Director of the Summit County Planning and Building Department, feels there need to be two separate approaches to planning in his culturally complex jurisdiction. The East County is conservative farming and ranching country backed up against the Uintas. There, traditional zoning is used, with 20-acre lots exempt from subdivision review. The West County is completely different.

Located only 35 miles from Salt Lake City, the West County region is being overrun by land speculators. In response, Summit County devised separate planning and zoning regulations for this broad area known as Snyderville Basin. There, only parcels 40 acres and above are exempt from planning board review. Even in such cases, the issuance of building permits is used to ensure that developments comply with zoning district criteria. For

The Wasatch Back-Valleys

large subdivision proposals, a point system is used to evaluate each project's design and impacts. The Snyderville Basin Development Code recognizes the importance of protecting open lands.

> *An area of almost complete consensus among local residents is the need for plenty of open space. The [resulting] density bonus provisions allow developers willing to provide significant open space to use their properties more intensively than those developers who choose to meet only the minimum requirements of the code. (Snyderville Basin Development Code, 5-23)*

The density bonus is invoked when at least 25 percent of a project area is set aside as open space. Then a complex point system is used to calculate allowable increases in development density on the remaining land. Criteria include such things as slope steepness, soil type, proximity to commercial areas and ski resorts, and use of solar or other alternative energy.

Even such far-sighted regulations have done little to slow the land rush. Developments such as Elk Park, Timberline, Summit Park, and Parleys Summit cover former ranchlands. Some 1,000 summer home lots have been platted up a single dirt road west of Coalville in fireprone oak brushland. Evacuation would be impossible if the roadway were blocked by a wall of fire. The Pinebrook project has received master plan approval to build 2,150 housing units. However, it is the Jeremy Ranch debacle which best illustrates the downside of the aggressive myopia of many realtors.

The Jeremy Ranch was an ambitious "new town" project which was to extend over some 15 square miles. The initial phase resulted in construction of an Arnold Palmer–designed golf course, handsome "spec" houses, and miles of hiking, cross-country skiing, and riding trails. Financial backers imagined this first phase would "build out" at 1,750 units, and a permit was received sanctioning this grandiose master plan. However, financial problems struck early, as few lots were purchased. The expensive buy-in cost, high Special Improvement District (SID) taxes, golf course membership fees, and the often icy winter commute to Salt Lake proved to be terminal liabilities. The area was even seen by the public as too far from the Park West, Park City, and Deer Valley ski resorts lying to the south. A last-ditch effort was made to sell land via satellite, but ultimately the project went bankrupt. The Virginia Beach Federal Savings Bank repossessed the property in order to retrieve some of the $32.7 million it was owed. Ever since, financially strapped Summit County has been understandably reluctant to issue SID bonds for even worthy developments.

PARK CITY

Park City, Utah, is where California's George Hearst began to build the family fortune. In 1872, he bought the Ontario silver mine for $27,000. It was a steal. The vein lode ores and bedded replacement deposits of quartz, pyrite, calcite, galena, sphalerite, and tetrahedrite eventually yielded some $50 million. Using this profit, Hearst went on to develop the Homestake Gold Mine in the Black Hills of South Dakota and the Cerro de Pasco Copper Mine in Peru. By 1880, a stampede of Irish, Cornish, English, Scot, Scandinavian, and Chinese miners swelled Park City to a population of 3,500.

Amid fortunes made and lost, the story of Susanna Bransford Emery Holmes Delitch Engalicheff stands out. She was a shrewd businesswoman and international bon vivant, better known as simply "The Silver Queen." During the 1880s, she earned over $1,000 a day from her stockholdings in various mines tunneled beneath Park City. She later parlayed this into a $200 million fortune and built a Park City mansion called the Amelia Palace, known for its elegant chandeliers and pure silver doorknobs. Cagey, diversified reinvestment brought the Silver Queen through the 1893 collapse of silver prices, and she continued to live in grand style, traveling the world until her death in 1942 at the age of 83. Park City didn't fare as well. In 1898, a great fire incinerated parts of the town.

The community remained in the economic doldrums until 1963, when the United Park City Mines Company built a ski resort. The company believed:

> *A once fabulous mining boom town can again become a bonanza for those with the vision to see its vast potential as a year round recreational and residential resort center . . . and therein lies the "silver lining" to the dark economic cloud which has hung over Park City. (United Park City Mines Company, 1)*

That same year, the main street of the city was paved for the first time and a new water system and sewage facilities were built. Lots in Park City were sold for back taxes—a mere $25.

In 1968, California developer Robert Major built the Park West ski area

The Wasatch Back-Valleys

3 miles from town. By 1970, city lots were fetching $2,500 each. The Park City master plan of that period urged that downtown buildings which were being converted to businesses retain their "old mining town flavor." The "rustic hip" theme caught on. Demand for skiing increased so much that a ski resort named Deer Valley was opened in 1981.

Park City is today the Hong Kong of Utah. It is an insular, expensive recreational and retailing colony surrounded by a vastly different culture. Superb skiing attracts the wealthy to a state which would otherwise hold no intrigue for them. This exclusive clientele is reflected in Park City's hotels, with names such as Chateau Apres, the Innsbruck, Stein Ericksen Lodge, and La Concierge. It is possible to dine on Tchiban sushi, Hunan Chinese, Texas barbecue, blackened snapper, creole gumbo, pasta primavera, or poached salmon only a few miles from the chicken-fried steak cafes of Coalville. Park City even has its own brewery, which produces a local favorite known as Wasatch Ale. An invisible but insurmountable economic divide separates the boutiques, galleries, malls, and eateries along Park Avenue from the sheep ranchers who live out in the sagebrush to the east.

A swirl of condominium and single-family home developments encircles Park City. The Enclave, The Aerie, Solamere, American Flag, and other projects have been built on the oak and aspen hillsides east of town. Park Meadows, Holiday Ranch, Ridgeview, Prospector Village, Willow Bend, Quail Meadow, Park West Village, and Meadow Wild spread out over pasturelands and sagebrush grasslands north along Route 224. Road signs within this mosaic aim to evoke a certain elite profile. Royal Street, Supreme Court, Pinnacle Drive, Aerie Drive, Payday Drive, Golden Way, Mellow Mountain Road, and Kings Court are a few of these self-consciously prestigious addresses.

The Deer Valley development is a well-planned cluster of earth-tone condos with stone chimneys and passive solar designs. A park with artificial ponds and a gazebo lies at its center. The roads are superb, the pedestrian paths well-lighted, and the utilities have all been buried underground. The layout mirrors natural slope breaks to create a visually unobtrusive pattern. Deer Valley is a monument to good planning. Sensible in terms of services (buses run through the neighborhood), dense to minimize land requirements, and close to ski lifts, it is a textbook example of how such projects should be done. However, a systematic open space conservation effort is only beginning to be incorporated into the planning process here and throughout the Park City area.

In the late 1980s, the Park City Planning Commission held meetings with representatives of the Trust for Public Land to look into land conservation options. From that meeting came ideas for a rails-to-trails project on a 30-mile stretch of Union Pacific track running east from the city and an open space corridor along Route 224. Yet to date, only one project has actually been completed.

In 1988, Park City planners and the Equity Properties Corporation reached an annexation agreement with a strong open space component. The document stipulated that 191 acres of the 278-acre property would remain undeveloped and that 116 housing units could be built on the 87-acre remainder. However, Park City and the developers fell into a familiar trap regarding conservation easements. Part of the agreement stated:

> *The dedication will be intended to permit Equity to receive whatever tax benefits may be possible from this open space dedication. The dedication is not being required by Park City but if made will be made by Equity on its own volition.*

However, on the same page the document states:

> *Park City will waive its Parks and Recreation fees in the amount of $129,375* in consideration of *[my emphasis] the preservation of the meadow area in open space.*

Stating this *quid pro quo* jeopardizes the charitable intent of the donor. In the eyes of the IRS, the fee waiver will reduce the value of the conservation easement deduction by exactly $129,375. Any remaining development value which has been foregone would then be fully deductible as a charitable gift. This appears to be a case where, in the absence of land trust assistance, a land protection technique was not applied with sufficient care.

Park City is presently negotiating with the owner of a 200-acre dairy farm along Route 224. The planning staff grasps the potential of the conservation real estate concept, but the approach is not fully embraced. The planning system in Park City is already the most sophisticated and therefore the most time-consuming in Utah. A land conservation program is largely seen as yet another complex burden, one best left to others to pursue.

In response, in 1990 local conservationists formed the Summit Land Trust—the first and only such group in Utah. The mission of the trust is to preserve lands of scenic, agricultural, historic, and ecological importance throughout Summit County. After receiving a start-up grant from the Eccles Foundation, the trust has hired a director, begun a county open space planning effort, and completed some outstanding projects. A 26-mile-long recreational trail has been completed on an abandoned railroad right-of-way between Park City and Coalville—the first "rails to trails" conversion in the state. In late 1992, the trust completed its first conservation easement on 220 acres of ranchland. In addition, the group has been actively involved in environmental education. This includes a program that encourages school kids to write essays on their vision of what the community has been and what it can be through enlightened stewardship of the environment.

Summit County is the only setting in Utah where voluntary approaches to landscape conservation are beginning to be accepted. Given Park City's somewhat atypical character compared to the rest of archconservative Utah, it isn't surprising that land trusts have taken root here first. In the rest of the state, the challenge will likely prove much more difficult. But, at least the first step has been taken.

SANPETE COUNTY

The classic agrarian Mormon landscape is perhaps nowhere better seen than in Sanpete County. Lying between the San Pitch Mountains and the Wasatch Plateau, this pastoral valley-based Zion still contains the traditional rural signature of Mormon life. Sanpete is Wallace Stegner's "Mormon Country." The wide streets of four-square platted towns still are paralleled by irrigation ditches. Barns, outbuildings, and farmhouses remain concentrated within the townscape. The LDS ward continues to be the focal point of people's religious and social lives. Farming and ranching still dominate the county's land-use pattern and economy. So remarkable is the place that one of the finest works of geography in recent years has been written about it, Gary Peterson and Lowell Bennion's *Sanpete Scenes: A Guide to Utah's Heart*. Regrettably, the very landscape and historic structures which have drawn so much attention still lie essentially unprotected from incompatible development and the wrecker's ball.

Sanpete County has never had a land-use planner, land trust, or historic preservation program. The county attorney serves on the Planning Commission and is the nearest thing to a planning director in Sanpete. The county surveyor deals with building codes and zoning. A government official confides, "We'd be better off with a planner, but we don't have the money."

Utah: An Elusive Zion

Land development activity has thus far been focused in the mountain canyons and foothills at the north end of the county near Indianola. Most subdivision is to provide recreation homesites for Provo residents. Despite countywide zoning which limits housing to mostly 5- to 10-acre tracts, projects are essentially unaltered by the planning process. Local review consists of an evaluation of the quality of road construction and sewer/water systems. In fact, Sanpete County demands that developers sign agreements which relieve the Road Department of any responsibility to maintain or plow new subdivision roads. According to Planning Commission members, their main intention is to limit the county's financial obligation to serve developing areas. Provision of parks or conservation of open lands is not even mentioned in the dialogue.

In part, this may be because of local bitterness over purchase of big game habitat by the Utah Division of Wildlife Resources. The agency has assembled a 28,000-acre checkerboard of wildlife management areas along the sagebrush grassland/pinyon-juniper benchlands on the east side of the valley, extending from Spring City south to Centerfield. The county attorney notes that despite the animosity about the removal of these lands from the tax rolls, people still want to leave open the option of one day selling some ground to the state.

Sanpete County was once known as "Utah's Granary." However, today it isn't wheat or sheep which dominate the agricultural sector of the economy, it's turkeys. The Moroni Feed Company was formed in 1937 by a group of local turkey growers. Today this company is part of a regional co-op known as Norbest. Sanpete County is now one of America's top ten centers of turkey production. The grain which is fed to toms in long-pole barns is mostly hauled in from the Midwest. There are over one-hundred turkey growers in the county. Each outfit raises between 40,000 and 120,000 birds. In 1987, some 4 million turkeys totaling 64 million pounds were produced. In 1992, the harvest topped 70 million pounds. The business is volatile, with episodic die-offs from poultry flu, sinusitis, and excessive heat or cold. The Jensen operation near Moroni lost 3,700 birds to a snowstorm in 1988. In 1976, 600 birds froze to death, and during the 1960s, the Jensens suffered heavy losses during prolonged hot spells. Despite all this, a local grower reveals a deep love for the work:

We produce about 850,000 pounds of turkey each year, and that's not too bad. When we get tired of taking care of the turkeys, we tend to the cattle and sheep or work in our grain and fields. It's a nice life.

The Wasatch Back-Valleys

In 1988, Norbest opened a huge cooperative turkey processing plant in Moroni. The town was a fine choice. Moroni is a quiet, small city with a pleasant mixture of brick and wood frame houses. Many have porches with two chairs set out next to tubs of red geraniums. American flags hang all over town. The Norbest facility sits out on Moroni's south side.

On the days just before Thanksgiving, things slow down a bit. Yet, the yard of the plant is still filled with long, flat turkey trailers lined up at the delivery dock. One driver explains that he has raised turkeys "all over Sanpete County for nearly all my life." The man's birds were destined for Christmas dinner feasts. "We're winding down the season now. They'll kill only about 20,000 birds today."

On one visit to the Norbest plant, a fortyish, blond supervisor burst from a door in the main building yelling, "Who are YOU! We don't allow any pictures!" When asked why, he admitted, "We've been having trouble with animal rights activists taking pictures and condemning our industry. We kill turkeys as humanely as possible, and we're proud of the job we do." It seemed true. Norbest looks like a well-run outfit. After explaining how turkeys are slaughtered in New England using metal funnels to hold the birds' heads as their throats are slit above buckets to catch the blood, which is later spread on the fields, the supervisor was at least partially convinced no one present was a vegan terrorist. He walked away saying, "No more pictures, *PLEASE!*"

No one likes to see birds killed, but there are trade-offs which make it worthwhile. Turkeys raised in Sanpete County provide superb low cholesterol meat. Also, the health of the turkey industry is the main reason so much of the county's farmland, open space, and wildlife habitats remain intact. Just before leaving, I bought a Norbest turkey at a Moroni market and placed it in the bed of the pickup, packed in Sanpete snow.

Jerusalem is a typical rural hamlet in the Sanpete poultry landscape. The sides of the roads are often covered with a dusting of white feathers. Jerusalem rests up a partially overgrown road cratered with potholes. The town is a mixture of old farm houses, collapsed wooden barns, 1950s vintage suburban homes, and metal-roofed turkey coops. Red-tailed hawks perch in the cottonwoods watching pheasants dart in and out of the tall pasture grass. To the north, the 11,877-foot snowy summit of Mount Nebo stands like a lighthouse at the end of the Utah Rockies.

In the adjacent hamlet of Freedom, the scene is repeated. Pens full of huge tom turkeys assume their comical threat poses. Yard bosses put their heads down, raise their wings, and begin waddling side to side before schlepping forward, gobbling loudly, and glaring through dark, stupid eyes.

The cemetery lies west of the coops up Freedom Lane. The name Draper or Taylor has been carved on nearly every headstone. Freedom is a town where people don't remarry after a spouse dies. Several stones have the image of the LDS Manti Temple on them. A seemingly disproportionate number of the graves are filled with men who died in America's wars. This little cemetery in farming country touchingly marks the passing of generations, but not of families.

Up the road a way is Spring City, an archetypal four-square Mormon community of 715 located in the center of Sanpete County. In 1981, the entire city was listed on the National Register of Historic Places. Spring City spatially documents Mormon city planning and contains many remarkably well preserved farm buildings and homes which predate World War I. The Romanesque Revival LDS Tabernacle was built in 1914 of native, buff-colored, oolitic limestone blocks. The turn-of-the-century Spring City Public School was constructed of locally fired brick. Throughout the city plat are adobe, brick, stone, wood frame, and log structures. Architectural styles include Greek Revival, Queen Anne, Victorian Eclectic, vernacular, and design hybrids. Many of the city's oldest buildings have been torn down. However, much of what makes Spring City extraordinary remains. Unfortunately, the National Register classification does not assure this will always be so. Register status simply makes funds available on a matching, competitive basis through the State Historic Preservation Office for sensitive restoration of buildings. Any or all of the privately owned structures in the city district could be razed at any time.

In 1852, Brigham Young chose the site for Spring City and instructed James Allred on how the plat should be drawn. All buildings were to be in town arranged on streets numbered North, South, East, and West of the Tabernacle. The outlying valley was to remain as open pastures, grainfields, and wetlands. Today, around Spring City and the rest of Sanpete County, the Prophet's vision has held its ground. In fact, Brigham Young has been the only land-use planner that Spring City and Sanpete County have ever known.

That has been good enough until now. The valley has been discovered as a retreat, and Uinitee Real Estate of Mount Pleasant is beginning to sell a great deal of land. Its business sign blares: "Recreation Center—Mountain Ground" at passersby. Prices are cheap. Ten acres with aspens and stream frontage can be bought for $12,000. A three-bedroom brick home on a half-acre lot is advertised for $15,500. One hundred acres is offered at $29,000. These land values are drawing more and more people from Provo, Nephi, Richfield, and beyond.

The Mormon agricultural landscape has been devoured by urbanization throughout much of Utah. Sanpete County is the best of what remains of that cultural legacy. Preservation of the buildings of Spring City or other towns without conservation of the land which embraces and gives meaning to them would be a Pyrrhic victory. Sanpete County merits recognition as the setting for a different kind of stewardship in America—a cultural national park. The New Jersey Pine Barrens and New York's Adirondack Park are working examples of our attempts to follow the European vision of a national park as a sustainably occupied landscape.

The key to the creation of this working park would be cooperation among local Sanpeters, the LDS Church, the county government, state and Federal agencies, the governor, and the Utah congressional delegation. From this exchange of ideas could emerge an equitable and wise plan for this lived-in landscape. Life would continue basically as it is. The point of the park would be to eliminate only those land-use choices which would be destructive of the valley's natural and cultural environment. Bed and breakfast farms could be set up to take in guests and market local crafts and agricultural products. Land subdivision and development would have to be tightly controlled or prohibited. Conservation easements, land exchanges, or monetary payments would be needed to compensate landowners for giving up their subdivision rights. Farmers and ranchers would benefit by being paid the development portion of their land's value while still owning and managing it as before. A local land trust could expedite the necessary transactions. The LDS Church might use the park as a place to showcase the historic roots of the faith and the values Mormonism espouses. School children from all over Utah could be bused in for field trips to milk cows, feed turkeys, ride horses, and observe deer, elk, and other wildlife. With 85 percent of Utah's residents now urban dwellers, Sanpete could offer them an opportunity to again touch the land and their historic roots.

Utah already attracts millions of visitors to its five existing national parks. The addition of Sanpete National Cultural Park would offer these tourists a fascinatingly different Zion. The logistics of creating such a park are daunting. Local resistance would be stout. However, the economic benefits to Sanpeters could be enormous, and the opportunity to protect a landscape as rare and full of meaning as this one makes this a challenge worth facing.

20

IF YOU
BUILD IT,
HE WILL COME

By the rivers of Babylon
There we sat down, yea, we wept
When we remembered Zion.
—PSALMS 137 : 1 (KJV)

Utah is deceiving. On the surface, the culture bears a striking similarity to
the rest of the country's mainstream, conservative way of life. Once their
agrarian, socialistic, utopian model began to fall away, Mormons, led by
church leaders, embraced the central message of American capitalism, redefined it, and brought it into the fold of their faith. The drive to "Build Up
Zion" lies at the center of Mormon religious instruction. However, trans-
forming a gloriously chaotic natural world into sacred order has worked far
better as theology than ecology. Because of their shared dependence on destructive exploitation of the natural environment, Mormon Millennialism
and American Manifest Destiny have been essentially similar in effect.

When Zion was to be a landscape of patterned agricultural villages laid
out according to Joseph Smith's divine plat, the scale was still small enough
to limit environmental losses to nodes in the canyons and benchlands around
Salt Lake City. However, when an urban way of life gained dominance, the
environmental consequences became widespread.

While the isolation Brigham Young sought in the desert worked in a
social sense, it presented the pioneers and future generations with unforeseen
environmental complexities. Aridity may have been conducive to piety, but
it has been a bane to logic. Mormons have never seen water as a limiting
factor for population growth and economic expansion. Although Joseph
Smith spoke of stewardship, Brigham Young was in charge of actual settlement. And Young, the Great Colonizer, more than anything else was a stern,

∧∧∧∧∧

hands-on CEO who ran the Church as a real estate development corporation. This tradition still guides Utah life.

In Mormon history, "Zion" has served a plurality of purposes ranging from theological abstractions to geographic certainties. The word has expanded, contracted, and changed shape and meaning depending on the needs of the Church. The strongly centralized LDS Church, long resistant to change, engages in regular revisionism regarding the meaning of Zion. In Mormon theology and culture, Zion has been a placeless heaven, an array of places from New England to Utah, a mountain, valleys, the desert, a City of God, agrarian, industrial, peaceful, warlike, all Mormon people, a spirit, a state of grace, a nation (Deseret), a vision, a journey, a faith, a force, the celestial realm of Heaven, economic security, all LDS temples, the entire earth, many worlds, and the Universe. It has been an all-embracing term symbolizing an ideal existence.

The search for the ideal has carried the Mormons across thousands of miles and through a broad expanse of philosophical terrain. Zion has been elusive. First it was one way of life and then another. Often there was little official recognition that a shift in cultural direction had occurred. As visions of an impending Millennium grew stronger or diminished, the Church repeatedly redefined the Faith and the Way. What is traditionally spoken of as "The Mormon Landscape" is a nineteenth-century agricultural pattern based on Joseph Smith's utopian religious view which reflected the era when the Church was formed. Isolated in Utah, with roots in New England and a knowledge of other traditions, the Mormons, through a kind of vicariance cultural geography, created a distinctive way of village life focused on their church and irrigated agriculture. However, outside of certain sheltered pockets such as Sanpete County, this archetypal pattern has largely disappeared.

The new Mormon landscape, the new Zion, is an urban sprawl reminiscent of scores of crowded corridors across the country. The village pattern arose during a period of divergence and differentiation. Over the years, improved transportation, mass marketing, and the advent of telecommunications have caused much of Utah's vernacular distinctiveness to be supplanted by a generic sameness.

In the past, the four-square platted Mormon village was a spatial signal to God that the people were living righteously. They lived close together in gridded sanctuaries with LDS chapels at their hearts. The landscape itself demonstrated where and how to live. It served to reduce ambiguity about what was expected. The mountains were to be exploited, and the valleys were to be sustainably farmed. This environmental partitioning served as the model for all the Church's colonization drives and possessed great religious meaning.

In the Mojave Desert, Indian people once placed stones in patterns to create outlines of animals. These intaglios may have been messages to the Great Spirit saying, "We believe. Favor us with plentiful game." For early Mormons, an agricultural landscape of small towns was their intaglio.

The transition to an overwhelmingly urban way of life has wrought tremendous changes in the state. Some believe that Utah's drive toward economic development, the growth of the defense and aerospace industries, attendant air and water pollution problems, and farmland losses have weakened the Church's hold on people's lives. One interpretation of modern Utah is that it is simply Americanism writ large and without apology. That is untrue.

The new Mormon landscape is a variation on what St. Augustine called the City of God. Joseph Smith's Plat of the City of Zion was a design for small, distinct farming-based cities of no more than 20,000 people. However, the man in charge of settlement, Brigham Young, saw no such limits. The Wasatch Front is today a solid congestion of roads, houses, refineries, stores, and warehouses with over 1.3 million people. The failure of Mormon culture to embrace land conservation, its rapid acceptance of urban life, and its unwillingness to face the limits of water supply signify that the new intaglio is the megalopolis. Young's industrial pragmatism has now erased all but a small vestige of Smith's agrarian ideal.

For most Mormons, supporting a national defense bureaucracy, Utah's boundless economic expansion, creating large families, and accumulating personal wealth are today's chosen signals to God that they are true believers. Shopping malls and power plants have assumed the role once played by rural farming landscapes as symbolic messages to the Lord. Members of the LDS Church await the Millennium, when Jesus will come down to Earth (Utah) and rule as a divine King for a thousand years. To Mormons, the quest is to manifest a limitless, eternal city for God and to pursue this grand mission free from secular doubt. They listen rapturously to their leaders' unadorned promise—if you build it, He will come. And they believe.

In a 1978 address at Brigham Young University, Mormon scholar Hugh Hibley presented a surprisingly different message to the student body:

If You Build It, He Will Come

*Ever since the days of the Prophet Joseph Smith, presidents of the
Church have appealed to the Saints to be magnanimous and forbearing
toward all of God's creatures. . . . Man's dominion is a call to service,
not a license to exterminate. (May 24, 1978)*

Conservationists have found that this sentiment is not widely shared in
Utah. It is extremely difficult to make significant headway in land-use planning and conservation in a state where the majority of the people await a
Millennial rescue when God will come to Utah and "perfect it as their paradisiacal home."

Utah lacks what Colorado has in abundance—land trusts. While Colorado has 27 trusts, Utah has only one, the Summit Land Trust in Park City.
Except for Salt Lake City's sporadic efforts, local government open space
programs are also essentially nonexistent. The Nature Conservancy has protected 26,000 acres (25 projects) in the state. However, its mission is the
protection of threatened and endangered species, not saving open landscapes, big game habitats, and ecologically sound but relatively common
natural communities. Conservation of these lands should be the responsibility of land trusts and city/county governments. However, this sort of
landscape saving requires local people to initiate local action. Unlike their
counterparts in most of the country, residents of Utah have made few moves
to do so.

However, land conservation concepts are beginning to be more widely
diffused throughout the state. The Nature Conservancy is currently involved
in projects ranging from the Deep Creek Mountains to the Colorado Plateau.
TNC brought the LDS Church into an effort to protect the 462-acre Lytle
Ranch near St. George in southern Utah. Brigham Young University raised
the purchase price and now uses the biologically rich property as a research
station. Among the ranch's biota is the Virgin River spinedace (*Lepidomeda
mollispinis mollispinis*). This fish species is a candidate for Federal listing as
endangered and is known from only six other streams in the world. Such a
remarkable precedent could lead to more projects with the Church, with
ancillary benefits throughout the state. The creation of an LDS-sponsored
Mormon Trail Land Trust and the protection of the Sanpete County historic
landscape seem obvious places to start. However, it is worth reiterating that
while Colorado has the most land trusts of any state in the West, Utah has
the least. Until this major obstacle is overcome, landscape conservation really
doesn't stand much of a chance.

As in Colorado, the need for habitat and open space conservation is well accepted within Utah's land management agencies. The State Division of Wildlife Resources and the State Parks Department are doing some excellent work. Other bureaus and divisions are also achieving meaningful results. However, as in many aspects of Utah life, it would be a mistake to assume that the motivation of state employees mirrors the Gentile world.

A fine Mormon man who works in a position of substantial authority in one of the state's land management departments once expansively described how much he believed in environmental conservation. When asked why, he confided, "I feel that the Millennium is a real event that's coming soon. And when God comes to Utah, He'll bring with Him all the Lost Tribes and all the Mormons who have passed on. And you know," he said, "we're going to need the room."

Utah would be better off if more people believed as this man does. The state was the last in the Union to form a Department of Natural Resources. The world's most accomplished savers of genealogical records seem strangely unconcerned about historical buildings and the natural environment.

The synergy of belief in both Millennial bailouts and technological fixes is a potent force working against land conservation activities in the state. As a result, the urban model stands largely unchallenged. With recent conversion of the Central Utah Project to an almost entirely urban water supply delivery system, Utah has publicly admitted what has long been apparent. Growth for its own sake and growth to glorify God are the state's credos.

A drive from Brigham City south to Provo provides a transect through what has been recognized as the "core" of the Mormon culture region. Such a journey once traversed one of the most profoundly recognizable of all of America's "Wests." Today, the trip is much like driving through the Los Angeles Basin. The regimented control of the Uinta and Wasatch watersheds is reminiscent of Switzerland. The dominance of one faith amid a huge military buildup evokes Israel.

Utah remains a unique place, but much of it no longer conforms to our long-standing mental map. Growth, the universal solvent of cultural variety, has erased much of what Mormons and Gentiles loved about the landscape. In accepting that laissez-faire development serves God, Utahans have traded a rich legacy for an illusory affluence.

They are not alone in choosing this fool's bargain. What makes it so unfortunate is that a fine, environmentally based vision of life was once the centerpiece of the LDS Church. Mormons were to be enlightened stewards obliged to a divine landlord. There was a chance things would be different here. Instead, Mormon theology has made Utah the most arduous cultural terrain in the Rocky Mountain West for land conservationists.

On mid-December evenings, traffic is thick around Temple Square in downtown Salt Lake. Thousands of smiling people gather to hear the Mormon Tabernacle Choir sing Christmas songs. The domed Tabernacle fills, and latecomers stand out among the trees and shrubs ablaze with sparkling Christmas lights. These are family nights. On one such memorable evening, a father lovingly cradled one of his six children, a boy, as the toddler reached upward trying to touch the bright white star high above the Square's manger scene. Just then, as the peaceful crowd stood in the polluted air, everyone, choir and families, young and old, began to sing the Mormon people's signature hymn, "Come, Come Ye Saints." There, where the dream of stewardship was first tested and where the transformation of the landscape has been most profound, the Mormons sang with conviction:

> *Come, come ye Saints*
> *We'll find the place where God for us prepared*
> *Far away in the West,*
> *Where none shall come to hurt or make afraid;*
> *There the Saints will be blessed*
> *We'll make the air with music ring*
> *Shout praises to our God and King;*
> *Above the rest these words we'll tell—*
> *All is well!*
> *All is well!*

PART IV
HOME

^^^^^

21
THE
GEOGRAPHY OF HOME

Singularity is almost invariably a clue. The more featureless and commonplace the crime is, the more difficult it is to bring it home.
—SHERLOCK HOLMES

The West was born of transience and a childlike vision of freedom. Much of it has been cut, mined, beaten, shot, and sucked dry. Yet despite decades of exploitation, the Rocky Mountain region still contains tremendous biological and visual richness. In most areas, it remains possible to live in the open, close to the land, although the motivation for doing so is as diverse as the people who populate this startling country.

Little credit for the West's remaining spectacular landscapes should go to land-use planning agencies. In a 1934 article, Berkeley geographer Carl Sauer wrote:

> *The westward movement in American history gave rise to the real estate boom, made land the first commodity of the country and produced the salesman promoter. It was the latter rather than any public official who planned and directed the settlement of new lands. (Sauer, 233)*

Such realtors were not enlightened, public-spirited, environmental designers. Most were driven by a quest for profit which spawned a sea of ill-conceived, ecologically damaging developments. Given the preeminent and unwavering power of the marketplace, open space still exists mostly because no buyer can be found or because the landowner chooses not to develop.

During the twentieth century, land-use planners have fabricated successive generations of rules and regulations, but land-splitting has continued

∧∧∧∧∧

largely unimpeded. Most planners perceive "open space" as developed parks or lands lying on the periphery of their zoning authority. The planning profession, as a whole, remains tragically uninformed about new methods of land protection. The destruction of "place" is now so institutionalized and bears such a veneer of normality or even inevitability that to challenge it is to appear out of step with the "real" world.

To paraphrase Sherlock Holmes's conundrum, a crime which is neither singular nor uncommon becomes acceptable. All the people in the land subdivision system—from the planner, banker, and county commissioner to the landowners and their neighbors—have rational-sounding arguments for their behavior. These people are not evil or even unlikable. The "problem" snaps into focus only when we stand back from the entire process to envision its likely outcome. Then, perhaps too late, we see as Haeckel did: "All our means are sane—our motive and object mad."

In the Rockies, a tradition of land speculation, homesteading, and development chicanery has been transmitted whole into the 1990s. Between 75 percent and 95 percent of all subdivision parcels created in the region are platted and sold without evaluation of the environmental, social, and economic impacts of such actions. The drive to evade government control has turned land-splitting into a vernacular craft with deep cultural roots. Subdivision exemptions are not loopholes or oversights, but purely intentional statements by Western legislators that the government has no business telling landowners what to do with their land. As a result, the subdivision laws of Western states function as *de facto* homestead acts.

This is the single most powerful driving mechanism contributing to land-use planning conflicts in the Rockies. To many residents of the region, Moses came down from the mountaintop carrying ten commandments and a survey rod. Most local planning systems have not fully addressed this attitude by adopting new methods. Many planners continue to hope that the public will "come around" if enough hearings are held and the perfect comprehensive plan and land-use regulations are written. The West's disappearing open landscapes reveal quite another outcome.

However, there are grounds for hope. Across the United States, 889 land trusts and scores of national groups are using an increasingly sophisticated array of conservation tools to protect geographic diversity. These private organizations, which have conserved over 7 million acres of private land, are now joined by numerous state and local governments which are raising public funds to acquire land and conservation easements. The purchase of development rights programs which exist in nine states have conserved 205,000 acres at a cost of $400 million (Daniels, 421–431). In 1988, California voters passed a $776 million statewide open space acquisition bond. Pennsylvanians have approved a $100 million bond to protect farmland around the Philadelphia urban fringes. New Jersey residents have passed a $300 million bond and Wisconsinites a $250 million bond to conserve key land. Floridians have raised $3.2 billion. Clearly, the message is that voluntary, compensating programs are necessary if liveable landscapes are going to exist in the future.

In the Rocky Mountain West, acceptance of the idea of conservation, by any means, has been strikingly uneven. There are 43 local and regional land trusts throughout the Rocky Mountain states. These groups have conserved a total of 171,760 acres of key habitats, farmland, and scenic open space. The Nature Conservancy has protected another 2 million acres, a great deal of this in New Mexico with projects such as the 321,000-acre Gray Ranch acquisition. The Montana-based Rocky Mountain Elk Foundation has conserved nearly 60,000 acres nationwide with the majority of its preserves in the West. Outside of Colorado, only a handful of city and county planning authorities in the Rockies have made any serious attempts to save their home lands.

Montana has 5 trusts and 100,106 acres permanently protected. The Montana Land Reliance has 96,000 acres under conservation easements. New Mexico's 5 land trusts have protected 19,306 acres; Santa Fe's Forest Trust has secured all of this total except the 145 acres conserved by the Taos Land Trust.

The 3 land trusts in Idaho have kept 2,466 acres in open space. In Wyoming, the Jackson Hole Land Trust has conserved 6,839 acres of private land. The state's other trust owns 20 acres and helped create a 5-mile-long greenway on the North Platte River through Casper.

Four trusts account for 151,995 acres, or 89 percent, of the Rocky Mountain regional total: the Montana Land Reliance, Colorado Open Lands, the Forest Trust (New Mexico), and the Jackson Hole Land Trust (Wyoming). Thus far, most trusts are not doing nearly as well.

Land trust work requires staff, funding, training, experience, and time. In the coming years, many youthful trusts are likely to substantially improve their performance. Others may wither away and disband, only to be replaced by fresh advocates of land-saving. The acreage totals, activities, and names of individual trusts change rapidly. However, the underlying historical and cultural geography of places will continue to exert a tremendous influence on the degree to which landscape conservation is accepted. And nowhere in the Rockies is what might be called "the principle of first effective land-use ideology" more forcefully revealed than in the stunning disparity between Colorado and Utah.

COLORADO AND UTAH

This contrast between the number of trusts in Colorado and Utah is curious given that on the surface the two states are so much alike. Both are characterized by recent, rapid gains in their mostly Anglo populations, by sprawling corridors of instant cities, by intensive industrialization focused on aerospace and defense industries, by massive transbasin water transfer systems, and by renowned ski resorts, national parks, and spectacular scenery. Although Colorado has 3.3 million people and Utah only half that, their populations are behemoth compared to other states in the region. Both Denver and Salt Lake City serve as transportation hubs and both are aggressively engaged in successful promotional campaigns designed to attract new businesses and residents. However, despite these parallels, Colorado and Utah are polar opposites regarding stewardship ideals. If Colorado is an ecotopia, Utah is a theotopia.

In Colorado, 27 land trusts, The Nature Conservancy (TNC), the Trust for Public Land (TPL), state agencies, and various city and county governments are engaged in scores of land protection projects. Colorado has the most balanced spread of land conservation accomplishments in the West. The state's land trusts have, despite long odds, protected 42,873 acres of prime habitat and open space. Colorado Open Lands has achieved notable successes at the Mount Evans Ranch and along Boulder Creek. The state's city and county open space efforts are the most successful in the Rocky Mountain West. The City of Boulder, Boulder County, and Jefferson County open space programs alone have functioned as land trusts to protect over 72,000 acres.

Utah has the smallest acreage total of conserved private lands in the Rockies. It also has the least number of land trusts, and the fewest acres protected by TNC. Utah's first land trust, the Summit Land Trust, has created a 26-mile "rails to trails" recreational corridor and received a conservation easement on 220 acres of open space. Surprisingly, the conservation programs of Utah's local governments have been moderately successful. This is deceptive, since 17,000 acres are in Salt Lake City's Wasatch Range municipal watersheds. All such acquisitions, except those within the recent City Creek Park and scattered hillside tracts, occurred decades ago because of pragmatic concerns over water supply, not because of ecological or aesthetic sensitivity. However, Utah's Jordan River State Park and the land purchase programs of its state agencies are commendable exceptions.

The striking divergence between Colorado and Utah cannot be fully explained by differences in political economy. These states are really separate "nations" with profoundly different belief systems. Colorado was founded on and continues to be a bastion of secular capitalism. Interests in land have always been exchanged without passing through a multifaceted religious prism as they do in Utah. While the development history of both states has been based on promotion and land disbursal, Coloradans have never shared a societal vision of the "correct" way of using the landscape or a vision of an afterlife. While Utah's Mormons strove to fulfill first an agrarian and then an urban ideal, Coloradans have been unfettered by a dominant religious outlook. Profit was their byword, not the word of the Prophet. Therefore, as environmental consciousness has grown in Colorado, the businesslike approach of land trusts has been widely accepted as simply a new, more responsible way of conducting commerce. Environmentalism seems to thrive in the Centennial State within the spiritual vacuum which is filled by Mormonism in Utah. While the difficulty of the task should not be underestimated, the mercantile engine which created Colorado now empowers an expanding conservation sector.

Growth has also been the prime directive in Utah. But in the Beehive State, it has long been seen as a way to worship God and to prove that the people are living righteously. Efforts to conserve landscapes encounter two interwoven and intensely powerful forces of opposition. Conservative beliefs about the inherent moral goodness of large families and unlimited economic development, together with a Millennial mandate to build Brigham Young's radiant *city* for God, have produced one of the most challenging settings in America for land conservationists.

If land trust advocates are to gain a foothold in the Beehive State, they must first gain an understanding of Mormon spiritual and secular attitudes about the highest and best use of the land. Since Mormons are generally extraordinary savers of genealogical records and are fascinated by their often inspiring history, perhaps a Church-sponsored land trust could be created to protect the lands traversed by the Mormon Trail or within such historic districts as rural Sanpete County, where the agrarian village pattern persists.

The Geography of Home

Such actions would be the logical first steps in tailoring the land trust concept to fit a unique set of cultural contours.

Growth itself is not necessarily a destructive process. However, as Edward Abbey often said, "Growth for its own sake is the ideology of the cancer cell." There is little effort made in Utah to plan where development will and will not go, let alone to discuss the environmental and social limits to growth. Although Coloradans have made efforts to address these fundamental concerns, overall, gains have been slow in coming. What results are the ghastly sprawls of the Wasatch Front and the Front Range. While one faces east and the other west, they both look out on tenuous futures.

CONSERVATION AND FREE WILL

Historian Patricia Limerick has written, "If Hollywood wanted to capture the emotional center of Western history, its movies would be about real estate" (Limerick, 55). She's right. Many Western "pictures" dealt with disputes between cowboys and Indians, cowboys and sheepherders, cowboys and bankers, cowboys and squatters, and so forth. These movies were about land tenure, and land tenure is about realty. Today, the preeminent Western conflict is between conservation and development interests. The decisions now being made about the landscape will shape not only the Earth, but the quality of our character.

Good geography is useful art. The practice of placecraft by land trusts is escalating nationwide, spurred on by the growing realization that progress leaves no retreat and that beautiful landscapes are as necessary as love for our well-being. Most governing bodies and Chambers of Commerce are gravely out of step with this trend. These officials often approach community development by promoting open-ended growth and creating industrial park sites which are somehow supposed to attract businesses and jobs to the area. These often vacant, weedy tracts are reminiscent of the sad little airstrips built by "cargo cults" in the jungles of Papua, New Guinea, in the post–World War II era. What is largely forgotten in this blind rush for business expansion is that a real sense of community comes from people's connection to each other and to the landscape.

Conservation is an opportunity offered by land trusts and open space programs; it is not a requirement. Inherent in this approach is an acknowledgement of private property rights and a willingness to compensate people for sacrifices they make. It seems a simple courtesy to thank landowners for being wise in actions which influence all our lives. It is this common sense which underlies landscape conservation work and makes it so philosophically appealing.

Conservationists are not stuffy antiquarians seeking to preserve phony examples of an imagined past. Land trust members come from all classes. Most see their work as what Montana writer William Kittredge called in *Owning It All*, "the pursuit of that absolute good luck which is some breathing time in a commodious place where the best that can be is right now" (99). *That* is what true community development is all about.

The struggle to salvage "Place" in an increasingly placeless world is a drama being played out in all the world's communities. Some of the rationales for conserving Amazonia are equally valid for protecting "the back forty" in the United States. Yet, attempts to reach environmental goals through excessive civil control will spawn further injustice and social unrest. Government programs which arbitrarily take ecologically significant lands away from people may seem to serve commendable goals for the protection of habitat, but in the long run they will be injurious to basic constitutional rights. Ultimately, an oppressive society and rich, beautiful landscapes are fundamentally incompatible. We must carefully walk the difficult divide between the common good and individual autonomy.

What seems best is a pluralism regarding places. It is worth remembering that magnificently unplanned, serendipitous landscapes are often the most fascinating and meaningful. They are diaries of our secret wishes and beliefs about how life should be lived. They can be mirrors of our spirit. Of course, some of these landscapes may also be horrifyingly unjust, ugly, and environmentally ransacked. Figuring out the difference is one way people in a community can learn about each other and find the common ground. Any lasting, positive change in our culture's attitudes about the land must come from such an exercise of free will. There is nothing more "American" than this.

CONCLUSION

The cultural landscape—the Earth as transformed by human hands—is the external expression of our search for salvation. Colorado and Utah are grand illustrations of how interior values shape geography. Those who advocate conservation across America simply must acknowledge the preeminent power of matters of the heart and spirit. As groups, Utah's Mormons and Colorado's

Gentiles *do* perceive the Earth differently. Yet noble, caring individuals reside on both sides of a seemingly impassible divide. While these states present a dramatic contrast, they also reveal a larger unifying theme. America's beautiful, environmentally healthy cultural landscapes, the places where we live, are eroding before our very eyes. And the responsibility to stop this preventable tragedy is ours. The techniques exist. It comes down to having enough courage and commitment to do the hard work which is required.

In parts of the country, people are inventing a permanence based on seemingly inconsequential changes. Like the statement of autonomy Joe Mondragon made by irrigating a long unused beanfield in John Nichols's book *The Milagro Beanfield War*, each conservation easement or other project stands as a reminder that a better future can be claimed one small step at a time.

Wallace Stegner has called Corvallis, Oregon, long a center of conservationists, one of the "seedbeds" where a new, permanent West will grow. More seedbeds are forming in other regions as land trusts bring together people who are grateful for the landscapes they inhabit. While this cultural shift has just begun, it is the harbinger of a fundamental redefinition of land and life in the country. Many people in the Rockies and elsewhere are awakening to the world around them with a deep thirst to know where and who they are. The searching has quickened to live in a true place, a home, where the voice of the land has not been silenced and to conserve that safe harbor so that we may remember real things too long put aside. This search often leads to our place of beginning—the geography of our childhood.

When I was ten, our family moved away from the only land I had ever known. We were leaving for the usual reasons people abandon economically depressed coastal towns in Maine. On that last strange and sad night in the big house, I collected five small personal treasures and carefully placed them in a blue coffee can. Sneaking out into the darkness, I walked by feel to a special spot in our woods, and by the beam of a penlight, dug a hole and buried the sealed can. I rolled a heavy boulder across the site as a marker.

Twenty-five years later I went back. The village hadn't changed much. The same large white houses with black shutters stood beside the same hayfields. Lanes Brook was still crawling with water striders. Along the Penobscot River, six rusted quonset huts and the old potato docks stood in foggy, diffused light. An old friend, Malcolm, put me up for the night and popped open a bottle of Moxie for me, like I was still a kid. We laughed

Home

about how the Red Sox still weren't really much good. Later, he made up a bed for me and laid out Sophie's red quilt at its foot for warmth.

The next afternoon, I headed up a familiar old gravel driveway and knocked on what was, by then, someone else's home for a quarter of a century. Our old friends recognized me at once and invited me in. The same wallpaper covered my room.

A half hour later I was out walking the land. My mother's mint garden scented the air. Upslope lay the overgrown field which was once waist-deep in our vegetables. At the field's far edge grew a rather average white pine with its base awash in a thicket of rose, hawthorn, and goldenrod. Instinctively, I climbed the tree before realizing I had used each foot- and handhold in precisely the same pattern as the last time I had sought its height—when Ike's golf score still made the news. Most of my treehouse had been brought down by years of Nor'easters blasting this unprotected forest margin. I must have meant business with the floor, though. It was still strong enough to trust, as I sat running my fingers over the wood. To the east, a freighter chugged upriver to Bangor.

After climbing down, I moved to the fringe of a small grove of trees. I recognized the clearing at once. The large boulder was rough to the touch and patterned in a mosaic of ruddy grains of feldspar, crystals of smokey quartz, and shiny flecks of black mica. It was still heavy. The first few shovelfuls brought only damp, brown soil, and underlying grayish till. I moved about a foot northward and dug again. Some reddish streaks—rust—and rotten metal. Digging with my bare hands, I moved aside stones, worms, and clay until I found myself holding one of the artifacts I was after, and then another, and another, until, finally, there were four. The fifth was not made of strong enough stuff. The land had simply absorbed it. I was stunned. I sat against a paper birch for a long time. The woods smelled like pine pitch and damp humus. The lily-of-the-valley my mother planted way out here had spread all over and was in full flower. My dad's scotch pines had grown thick trunks and long whorls of graceful branches. The quiet felt like an embrace.

I finally rose and handled each treasure a last time and placed them in the bottom of a red coffee can, paused to remove a single new amulet from the front pocket of my jacket, laid it with the others, and sealed the top down tight. Far off in the darkness a dog barked once as I shoveled earth into the hole and rolled the boulder back to where it belonged.

The Geography of Home

Land conservation is a lasting gift to your home. Each individual act of stewardship is a covenant with the Earth and the unborn—a personal ceremony which brings a kind of deliverance. A multitude of such acts will create a cultural change that assures we leave a rich legacy of landscapes to sustain both our children and the wild species of the world.

APPENDIX
LAND-SAVING GROUPS

NATIONAL CONSERVATION ORGANIZATIONS

American Farmland Trust
1920 North Street, NW, Suite 400
Washington, DC 20036
(202) 659–5170

Land Trust Alliance
1319 F Street NW, Suite 501
Washington, DC 20004-1106
(202) 638–4725

The Nature Conservancy (National Office)
1815 N. Lynn Street
Arlington, VA 22209
(703) 841–5300

The Nature Conservancy
Colorado Field Office
1244 Pine Street
Boulder, CO 80302
(303) 444–2950

The Nature Conservancy
Great Basin Field Office
P.O. Box 11486, Pioneer Station
Salt Lake City, UT 84110
(801) 531–0999

The Nature Conservancy
Idaho Field Office
P.O. Box 165
Sun Valley, ID 83353
(208) 726–3007

The Nature Conservancy
Montana Field Office
P.O. Box 258
Helena, Montana 59624
(406) 443–0303

The Nature Conservancy
New Mexico Field Office
107 Cienega
Santa Fe, NM 87501
(505) 988–3867

The Nature Conservancy
Wyoming Field Office
258 Main Street, Suite 200
Lander, WY 82520
(307) 332–2971

∧∧∧∧∧

Rocky Mountain Elk Foundation
P.O. Box 8249
Missoula, MT 59807
(406) 721–0010

Trust for Public Land (National Office)
116 New Montgomery Street, 4th Floor
San Francisco, CA 94105
(415) 495–4014

ROCKY MOUNTAIN LAND TRUSTS

Colorado

Aspen Center for Environmental Studies
P.O. Box 8777
Aspen, CO 81612
(303) 925–5756

Aspen Valley Land Trust
Box 940
Aspen, CO 81612
(303) 920–3806

Boulder County Land Trust
2635 Mapleton Ave., Suite 77
Boulder, CO 80304
(303) 447-1899

Boulder County Nature Association
P.O. Box 493
Boulder, CO 80306

Cherry Creek Trails and Open Space
Foundation
Box 454
Franktown, CO 80116
(303) 688–4075

Clear Creek Land Conservancy
650 Range View Trail
Golden, CO 80401
(303) 526–1151

Colorado Open Lands
1050 Walnut Street, #525
Boulder, CO 80302
(303) 443–7347

Continental Divide Land Trust
P.O. Box 588
Breckenridge, CO 80424
(303) 453–2901

Crested Butte Land Trust
P.O. Box 39
Crested Butte, CO 81224
(303) 349–5338

Denver Urban Gardens
227 South Grant Street
Denver, CO 80209
(303) 433–9585

Douglas County Land Conservancy
P.O. Box 462
Castle Rock, CO 80104
(303) 837–1917

Eagle County Land Conservancy
Eagle, CO

Estes Valley Land Trust
P.O. Box 663
Estes Park, CO 80517
(303) 441–3440

Jefferson County Nature Association
Lookout Mountain Open Space Nature
Center
910 Colorow Road
Golden, CO 80401
(303) 526–0594

La Plata Open Space Conservancy
P.O. Box 1651
Durango, CO 81302
(303) 259–3415

Mesa County Land Conservancy
P.O. Box 1246
Palisade, CO 81526
(303) 464–7214

Mountain Area Land Trust
Evergreen, CO (New group)

Palmer Foundation
P.O. Box 1281
Colorado Springs, CO 80901
(719) 576–4314

Poudre River Trust
Fort Collins, CO (Status uncertain)

Roaring Fork Land Conservancy
210 Airport Business Center, Suite PP
Aspen, CO 81611

South Suburban Park Foundation
6315 South University Boulevard
Littleton, CO 80121
(303) 837–0118

Southern Colorado Heritage Conservancy
P.O. Box 4407
Pueblo, CO 81003
(719) 547–0493

Southwest Land Alliance
P.O. Box 1066
Pagosa Springs, CO 81147

Telluride Land Conservancy
(303) 728–3757

Two Ponds Preservation Foundation
9190 Alkire Street
Arvada, CO 80005
(303) 424–0037

Wilderness Land Trust
765 Poplar Avenue
Boulder, CO 80304
(303) 440–7780

Yampa Valley Land Conservancy
Steamboat Springs, CO (Status uncertain)

Idaho

Boise River Trail Foundation
2733 Warm Springs
Boise, ID 83712
(208) 345–1657

Idaho Foundation for Parks and Lands
1020 West Franklin
Boise, ID 83702
(208) 344–7141

Idaho Heritage Trust
508 South Fifth Street
Boise, ID 83702
(208) 384–0176

Montana

Five Valleys Land Trust
P.O. Box 8953
Missoula, MT 59807

Flathead Land Trust
P.O. Box 1913
Kalispell, MT 59903
(406) 755–7000

Appendix: Land-saving Groups

Gallatin Valley Land Trust
P.O. Box 7021
Bozeman, MT 59771
(406) 587–8404

Montana Land Reliance
P.O. Box 355
Helena, MT 59624
(406) 443–7027

North Fork Land Trust
P.O. Box 11
Polebridge, MT 59928

New Mexico

Albuquerque Conservation Trust
121 Tijeras Northwest
Albuquerque, NM 87102
(505) 842–1525

Forest Trust
P.O. Box 519
Santa Fe, NM 87504
(505) 983–8992

Rancho Santa Fe Community Foundation
P.O. Box 811
Rancho Santa Fe, NM
(505) 756–1281

Taos Land Trust
P.O. Box 376
Taos, NM 87571
(505) 776–2281

Tonantzin Land Institute
P.O. Box 40182
Albuquerque, NM 87196
(505) 277–3653

Utah

Summit Land Trust
P.O. Box 1630
Park City, UT 84060
(801) 649–0220

Wyoming

Jackson Hole Land Trust
Box 2897
Jackson, WY 83001
(307) 733–4707

Platte River Parkway Trust
P.O. Box 1228
Casper, WY 82602
(307) 235–2807

SELECTED BIBLIOGRAPHY

PART I. LAND AND LIFE

Chapter 1. Geographers and Landscapes

Basso, Keith H. "Stalking with Stories: Names, Places, and Moral Narratives among the Western Apache." In Daniel Halpern, ed., *On Nature: Nature, Landscape, and Natural History*. San Francisco: North Point Press, 1987.

Brody, Hugh. *Maps and Dreams*. New York: Pantheon Books, 1982.

Jackson, J. B. *Discovering the Vernacular Landscape*. New Haven, Conn.: Yale University Press, 1984.

Leighly, John, ed. *Land and Life: A Selection from the Writings of Carl Ortwin Sauer*. Berkeley and Los Angeles: University of California Press, 1963.

Marsh, George Perkins. *Man and Nature; or Physical Geography as Modified by Human Action*. New York: Scribner, 1864.

Tuan, Yi-Fu. *Topophilia: A Study of Environmental Perception, Attitudes and Values*. Englewood Cliffs, N.J.: Prentice-Hall, 1974.

Wilson, Edward O. *Biophilia*. Cambridge, Mass., and London: Harvard University Press, 1984.

Chapter 2. Land Regulation and Conservation in America

Barrett, Thomas S., and Putnam Livermore. *The Conservation Easement in California*. Covelo, Calif.: Island Press, 1983.

Bosselman, Fred, et al. *The Taking Issue*. Washington, D.C.: Council on Environmental Quality, 1973.

Brenneman, Russell L. *Private Approaches to the Preservation of Open Land*. New London, Conn.: Conservation and Research Foundation, 1967.

———, and Sarah M. Bates. *Land-Saving Action*. Covelo, Calif.: Island Press, 1984.

Collins, Beryl Robichaud, and Emily W. B. Russell, eds. *Protecting the New Jersey Pinelands*. New Brunswick, N.J.: Rutgers University Press, 1988.

∧ ∧ ∧ ∧ ∧

Daniels, Thomas L. "The Purchase of Development Rights: Preserving Agricultural Land and Open Space." *Journal of the American Planning Association* 57, no. 4 (1991): 421–431.

Diehl, Janet, and Thomas S. Barrett. *The Conservation Easement Handbook*. Alexandria, Va.: Land Trust Exchange and the Trust for Public Land, 1988.

Eckholm, Erik. *Losing Ground: Environmental Stress and World Food Problems*. London: W. W. Norton, 1976.

Hoose, Philip. *Building an Ark: Tools for Preservation of Natural Diversity through Land Protection*. Covelo, Calif.: Island Press, 1980.

Institute for Community Economics. *The Community Land Trust Handbook*. Emmaus, Pa.: Rodale Press, 1982.

Land Trust Alliance. *1991–1992 National Directory of Conservation Land Trusts*. Washington, D.C.: Land Trust Alliance, 1991.

Levy, John M. *Contemporary Urban Planning*. Englewood Cliffs, N.J.: Prentice-Hall, 1991.

Little, Charles E. *Greenways for America*. Baltimore and London: Johns Hopkins University Press, 1990.

———. *Hope for the Land*. New Brunswick, N.J.: Rutgers University Press, 1992.

Mantell, Michael, Stephen F. Harper, and Luther Propst. *Creating Successful Communities: A Guidebook to Growth Management Strategies*. Covelo, Calif.: Island Press, 1990.

———, ———, and ———. *Resource Guide for Creating Successful Communities*. Covelo, Calif.: Island Press, 1990.

Montana Land Reliance and Land Trust Exchange. *Private Options: Tools and Concepts for Land Conservation*. Covelo, Calif.: Island Press, 1982.

Myers, Norman. *The Sinking Ark*. London: W. W. Norton, 1976.

Parsons, James J. "On 'Bioregionalism' and 'Watershed Consciousness.'" *The Professional Geographer* 37, no. 1 (1985): 1–6.

Sargent, Frederic O., et al. *Rural Environmental Planning for Sustainable Communities*. Covelo, Calif.: Island Press, 1991.

Small, Steven J. *The Federal Tax Law of Conservation Easements*. Washington, D.C.: Land Trust Exchange, 1986.

Stokes, Samuel N., A. Elizabeth Watson, et al. *Saving America's Countryside: A Guide to Rural Conservation*. National Trust for Historic Preservation. Baltimore and London: Johns Hopkins University Press, 1989.

Whyte, William H. *The Last Landscape*. Garden City, N.Y.: Doubleday, 1968.

———. *Securing Open Space for Urban America: Conservation Easements*. Washington, D.C.: Urban Land Institute, 1959.

Wright, John B. "Land Trusts in the USA." *Land Use Policy* 9, no. 2 (1992): 83–86.

———, and Anita P. Miller. "Preservation of Agricultural Land and Open Space." *The Urban Lawyer* 23, no. 2 (1991): 821–844.

Chapter 3. The Rocky Mountain West: Searching Country

Bailey, L. H. *The Country Life Movement in the United States*. New York: Macmillan, 1911.

Brower, David. "David R. Brower—Environmentalist, Activist, Publicist, and Prophet."

Sierra Club History Series. Berkeley: Regional Oral History Office, Bancroft Library, University of California, 1978.

Bugbee, Bruce A., and Associates. Inventory of Conservation Resources, Missoula County, Montana. Missoula, Mont.: Missoula County Commissioners, 1985.

Clawson, Marion. *Man and Land in the United States*. Lincoln: University of Nebraska Press, 1964.

Crawford, Stanley. *Mayordomo: Chronicle of an Acequia in Northern New Mexico*. Albuquerque: University of New Mexico Press, 1988.

deBuys, William. *Enchantment and Exploitation: The Life and Hard Times of a New Mexico Mountain Range*. Albuquerque: University of New Mexico Press, 1985.

DeVoto, Bernard. "The West: A Plundered Province." *Harper's Monthly Magazine*, August 1934, pp. 355–364.

Fradkin, Philip. *A River No More*. New York: Knopf, 1981.

Garreau, Joel. *The Nine Nations of North America*. New York: Avon Books, 1981.

Gregg, Josiah. *Commerce of the Prairies*. Lincoln: University of Nebraska Press, 1967.

Hayden, Ferdinand V. Report of the Commissioners of the General Land Office for the Year 1867. Washington, D.C.

Henderson, Robert E., and Amy O'Herren. "Winter Ranges for Elk and Deer: Victims of Uncontrolled Subdivisions?" *Western Wildlands*, Spring 1992, pp. 20–25.

Hibbard, Benjamin Horace. *A History of the Public Land Policies*. Madison and Milwaukee: University of Wisconsin Press, 1965.

Horgan, Paul. *Great River: The Rio Grande in North American History*. Austin: Texas Monthly Press, 1984.

Hunt, Charles B. Hunt. *Physiography of the United States*. San Francisco and London: W. H. Freeman and Company, 1967.

Layton, Stanford J. *To No Privileged Class*. Monograph No. 17. Provo, Utah: Brigham Young University, Charles Redd Center for Western Studies, 1988.

Limerick, Patricia Nelson. *The Legacy of Conquest: The Unbroken Past of the American West*. New York: W. W. Norton, 1988.

Meinig, Donald W. "American Wests: Preface to a Geographical Interpretation." *Annals of the Association of American Geographers*, June 1972, pp. 159–184.

———, ed. *The Interpretation of Ordinary Landscapes*. New York and Oxford: Oxford University Press, 1979.

———. *Southwest: Three Peoples in Geographical Change 1600–1970*. New York: Oxford University Press, 1971.

Powell, John Wesley. *Report on the Lands of the Arid Region of the United States*, facsimile of 1879 edition. Boston: Harvard Common Press, 1983.

Reisner, Marc. *Cadillac Desert: The American West and Its Disappearing Water*. New York: Viking Press, 1986.

Robertson, James Oliver. *American Myth, American Reality*. New York: Hill and Wang, 1980.

Smith, Henry Nash. *Virgin Land: The American West as a Symbol and Myth*. Cambridge, Mass.: Harvard University Press, 1978.

Stegner, Wallace. *The American West as Living Space*. Ann Arbor: University of Michigan Press, 1987.

————. In Foreword to A. B. Guthrie, *The Big Sky*. New York: Bantam Books, 1982.

Turner, Frederick Jackson. *The Frontier in American History*. Tucson: University of Arizona Press, 1986.

Westphall, Victor. *Mercedes Reales: Hispanic Land Grants of the Upper Rio Grande Region*. New Mexico Land Grant Series. Albuquerque: University of New Mexico Press, 1985.

Worster, Donald. *Rivers of Empire: Water Aridity and Growth of the American West*: New York: Pantheon Books, 1965.

Zelinsky, Wilbur. *The Cultural Geography of the United States*. Englewood Cliffs, N.J.: Prentice-Hall, 1973.

PART II. COLORADO: THE ICON

Abbot, Carl, Stephen J. Leonard, and David McComb. *Colorado: A History of the Centennial State*. Boulder: Colorado Associated University Press, 1982.

Armstrong, Ruth. *Promised Land: A History of the Sangre de Cristo Land Grant*. New York: Forbes, 1978.

Aspen/Pitkin County Planning Office. Down Valley Comprehensive Plan. Aspen, Colo., 1987.

Athearn, Robert G. *The Coloradans*. Albuquerque: University of New Mexico Press, 1976.

Barth, Gunther. *Instant Cities: Urbanization and the Rise of San Francisco and Denver*. Albuquerque: University of New Mexico Press, 1988.

Bird, Isabella. *A Lady's Life in the Rocky Mountains*. New York: G. P. Putnam's Sons, 1881.

Boulder County Comprehensive Plan. Boulder, Colo.: Boulder County Board of County Commissioners, 1986.

Boulder County Subdivision Regulations. Boulder, Colo.: Boulder County Board of County Commissioners, 1988.

Brown, Ralph. *Historical Geography of the United States*. New York: Harcourt, Brace & World, 1948.

Erickson, Kenneth A., and Albert W. Smith. *Atlas of Colorado*. Boulder: Colorado Associated University Press, 1985.

Fradkin, Philip. *Sagebrush Country: Land and the American West*. New York: Knopf, 1989.

Jefferson County Open Space Program and BRW, Inc. The Jefferson County Open Space Master Plan. Golden, Colo.: Jefferson County, 1989.

Land Trust Alliance. *1991–1992 National Directory of Conservation Land Trusts*. Washington, D.C.: Land Trust Alliance, 1991.

Lavender, David. *The Rockies*. Lincoln: University of Nebraska Press, 1981.

McHarg, Ian L. *Design with Nature*. Garden City, N.Y.: Doubleday, 1971.

McPhee, John. *Basin and Range*. New York: Farrar, Straus, and Giroux, 1980.

Nugent, Walter. "The People of the West Since 1890." In Gerald D. Nash and Richard W. Etulain, eds., *The Twentieth Century West*. Albuquerque: University of New Mexico Press, 1989.

Rocky Mountain Divide

Peirce, Neal R. *The Mountain States of America*. New York: W. W. Norton, 1972.

Rohrbough, Malcolm J. *Aspen: The History of a Silver Mining Town 1879–1893*. Oxford: Oxford University Press, 1986.

Sprague, Marshall. *Newport in the Rockies*. Denver: Sage Books, 1961.

Swartz, Virginia. "The How and Why of Conservation Trusts." *Colorado Outdoors*, November/December 1987, p. 10.

Thompson, Dr. Hunter S. *Songs of the Doomed: More Notes on the Death of the American Dream, Gonzo Papers Volume 3*. New York: Summit Books, 1990.

Webb, Walter Prescott. "The American West: Perpetual Mirage." *Harper's Magazine*, May 1957, pp. 25–31.

Whitney, Gleaves. *Colorado Front Range: A Landscape Divided*. Boulder, Colo.: Johnson Books, 1983.

Wiley, Peter, and Robert Gottlieb. *Empires in the Sun: The Rise of the New American West*. Tucson: University of Arizona Press, 1982.

PART III. UTAH: AN ELUSIVE ZION

Arrington, Leonard J. *Brigham Young: American Moses*. New York: Knopf, 1985.

——. *Great Basin Kingdom: An Economic History of the Latter-Day Saints, 1830–1900*. Lincoln: University of Nebraska Press, 1958.

——. "The Transcontinental Railroad and the Development of the West." *Utah Historical Quarterly* 37, no. 1 (1969): 15.

——, and Davis Bitton. *The Mormon Experience: A History of the Latter-Day Saints*. New York: Vantage Books, 1980.

Bancroft, Hubert Howe. *History of Utah: 1540–1886*. Las Vegas: Nevada Publications, 1982.

The Book of Mormon: Testimony of the Prophet Joseph Smith. Salt Lake City: Church of Jesus Christ of Latter-Day Saints, 1981.

Brodie, Fawn. *No Man Knows My History: The Life of Joseph Smith, the Mormon Prophet*. New York: Knopf, 1976.

Brown, Craig J. "The Allocation of Timber Resources within the Wasatch Mountains of Salt Lake County, Utah, 1847–1869." Master's thesis, University of Utah, 1982.

Bullock, Thomas. Pioneer Journal Manuscript. Brigham Young University Library, Provo, Utah.

Bureau of Reclamation. Central Utah Conservancy District, Jordanelle Dam and Reservoir. Orem, Utah, 1986.

Burton, Sir Richard. *The City of Saints and Across the Rocky Mountains to California*, facsimile of 1861 edition. New York: Knopf, 1963.

Campbell, Eugene E. *Establishing Zion: The Mormon Church in the American West 1847–1868*. Salt Lake City: Signature Books, 1988.

City Creek Master Plan. Salt Lake City: Salt Lake City Planning Office, 1986.

Clayton, James L. "An Unhallowed Gathering: The Impact of Defense Spending on Utah's Population Growth, 1940–1964." *Utah Historical Quarterly* 34, no. 2 (1966): 227–242.

Cottam, Walter P., and Fred A. Evans. "A Comparative Study of the Vegetation of Grazed and Ungrazed Canyons in the Wasatch Range, Utah." *Ecology* 26, no. 2 (1945): 145.

Doctrine and Covenants. Vol. 28. Salt Lake City: Church of Jesus Christ of Latter-Day Saints, n.d.

East Bench Master Plan. Salt Lake City: Salt Lake City Planning Office, 1987.

Federal Communications Commission. *Reports*, vol. 62, 2d series, p. 255. Washington, D.C., 1976.

Flores, Dan L. "Zion in Eden: Phases of the Environmental History of Utah." *Environmental Review* 7, no. 4 (1983): 325–344.

Fox, Feramorz Young. "The Mormon Land System." Ph.D. dissertation, Northwestern University, 1932.

Fradkin, Philip. *Sagebrush Country: Land and the American West*. New York: Knopf, 1989.

Francaviglia, Richard V. "The Mormon Landscape: Definition of an Image of the American West." *Proceedings of the Association of American Geographers*, 1970, pp. 59–61.

Fremont, John C. *The Exploring Expedition to the Rocky Mountains and to Oregon and North California*. Washington, D.C.: Henry Polkinhorn, 1845.

Gottlieb, Robert, and Peter Wiley. *American Saints: The Rise of Mormon Power*. San Diego, New York, and London: Harcourt, Brace, Jovanovich Publishers, 1986.

Haglund, Karl T., and Philip F. Notoriarri. *The Avenues of Salt Lake City*. Salt Lake City: Utah State Historical Society, 1980.

Hastings, Lansford. *The Emigrants Guide to Oregon and California*, reprint of 1845 edition. Princeton, N.J.: Princeton University Press, 1932.

Heinerman, John, and Anson Shupe. *The Mormon Corporate Empire*. Boston: Beacon Press, 1985.

Hinton, Wayne K. *Utah: Unusual Beginning to Unique Present*. Salt Lake City: Windsor Press, 1988.

Jackson, Richard. "Mormon Perception and Settlement." *Annals of the Association of American Geographers*, 1978, pp. 317–334.

———. "Myth and Reality: Environmental Perception of the Mormons, 1840–1865, an Historical Geosophy." Ph.D. dissertation, Clark University, 1970.

Journal of Discourses. Multi-volume set. Salt Lake City: LDS Church, n.d.

Kay, Jeanne, and Craig J. Brown. "Mormon Beliefs about Land and Natural Resources, 1847–1877." *Journal of Historical Geography* 11, no. 3 (1985).

———, and Dale L. Morgan. *The State of Deseret*. Logan: Utah State University and Utah Historical Society, 1987.

Linford, Lawrence L. "Establishing and Maintaining Land Ownership in Utah Prior to 1869." *Utah Historical Quarterly* 42, no. 2 (1974): 127.

Little, James A. "Biography of Lorenzo Dow Snow." *Utah Historical Quarterly* 14, no. 1 (1946): 98.

Matheson, Scott M. *Out of Balance*. Salt Lake City: Peregrine Smith Books, 1986.

Meinig, Donald W. "The Mormon Culture Region: Strategies and Patterns in the Geography of the American West, 1847–1964." *Annals of the Association of American Geographers* 55, no. 2 (1965): 191–220.

Merriam, Florence A. *My Summer in a Mormon Village*. Boston and New York: Houghton Mifflin, 1894.

Miller, David E. *Great Salt Lake Past and Present*. Salt Lake City: Publishers Press, 1987.

Morgan, Dale L. *The Great Salt Lake*. Lincoln and London: University of Nebraska Press, 1986.

Naifeh, Steven, and Gregory White Smith. *The Mormon Murders*. New York: New American Library, Onyx Books, 1988.

Nelson, Lowry. *The Mormon Village: A Pattern and Technique of Land Settlement*. Salt Lake City: University of Utah Press, 1952.

Peterson, Charles S. "Albert F. Potter's Wasatch Survey, 1902: A Beginning for Public Management of Natural Resources in Utah." *Utah Historical Quarterly* 39, no. 4 (1971).

Peterson, Gary B., and Lowell C. Bennion. *Sanpete Scenes: A Guide to Utah's Heart*. Eureka, Utah: Basin Plateau Press, 1987.

Quinn, D. Michael. *Early Mormonism and the Magic World View*. Salt Lake City: Signature Books, 1987.

Roberts, B. H. *History of the Church of Jesus Christ of Latter-Day Saints*. Salt Lake City: Church of Jesus Christ of Latter-Day Saints, 1930.

Roylance, Ward J. *Utah: A Guide to the State*. Salt Lake City: Guide to the State Foundation, 1982.

Smith, George Albert. *History of the Church of Jesus Christ of Latter-Day Saints*. Multivolume set. Salt Lake City: Deseret Book Co., 1950.

Snyderville Basin Development Code. Coalville, Utah: Summit County Planning Commission, 1988.

Stegner, Wallace. *The Gathering of Zion: The Story of the Mormon Trail*. Salt Lake City: Westwater Press, 1981.

———. *Mormon Country*. Lincoln and London: University of Nebraska Press, 1970.

United Park City Mines Company. "New Bonanza at Park City." In the Plan to Transform the Park City, Utah, Area. Park City, Utah, 1963.

U.S. Department of Interior. Bureau of Reclamation. "Central Utah Project Bonneville Unit: Municipal and Industrial System Summer Statement, 1979."

Utah State Planning Board. First Report on State Policies. Salt Lake City, 1935.

———. A State Plan for Utah. Salt Lake City, 1935.

Wasatch Canyons Preliminary Master Plan 1989–2009. Salt Lake City: Bear West Consulting Team, Salt Lake County Public Works Department Planning Division, 1988.

Woodruff, Wilford. Pioneer Journal Manuscript. Brigham Young University Library, Provo, Utah.

"The World According to Redford." *Outside*, December 1989, p. 44.

PART IV. HOME

Christian Science Monitor. "States Push Land Conservation." January 22, 1990.

Daniels, Thomas L. "The Purchase of Development Rights: Preserving Agricultural Land

and Open Space." *Journal of the American Planning Association* 57, no. 4 (1991): 421–431.

Kittredge, William. *Owning It All*. St. Paul, Minn.: Graywolf Press, 1987.

Limerick, Patricia Nelson. *The Legacy of Conquest: The Unbroken Past of the American West*. New York: W. W. Norton, 1988.

Lopez, Barry. "Mapping the Real Geography." *Harpers Magazine*, November 1989, pp. 21–24.

Nichols, John. *The Milagro Beanfield War*. New York: Holt, Rinehart and Winston, 1974.

Sauer, Carl O. "Preliminary Report to the Land Use Committee on Land Resources and Land Use in Relation to Public Policy." Edited by W. L. G. Joerg. Washington, D.C.: American Geographical Society, Science Advisory Board, 1934.

Stegner, Wallace. *The American West as Living Space*. Ann Arbor: University of Michigan Press, 1987.

Tuan, Yi-Fu. *"Thought and Landscape": The Interpretation of Ordinary Landscapes*. New York: Oxford University Press, 1987.

Vale, Thomas R., ed. *Progress against Growth: Daniel B. Luten on the American Landscape*. New York and London: Guilford Press, 1986.

Wright, John B. "Conserving Rocky Mountain Places: The Planning, Development, and Protection of Privately-Owned Lands in Colorado and Utah." Ph.D. dissertation, University of California, Berkeley, 1990.

INDEX

∧ ∧ ∧ ∧ ∧